Financial Engineering Explained

Series Editor
Wim Schoutens
Department of Mathematics
Katholieke Universiteit Leuven
Heverlee, Belgium

Financial Engineering Explained is a series of concise, practical guides to modern finance, focusing on key, technical areas of risk management and asset pricing. Written for practitioners, researchers and students, the series discusses a range of topics in a non-mathematical but highly intuitive way. Each self-contained volume is dedicated to a specific topic and offers a thorough introduction with all the necessary depth, but without too much technical ballast. Where applicable, theory is illustrated with real world examples, with special attention to the numerical implementation.

More information about this series at
http://www.springer.com/series/14984

Karel in 't Hout

Numerical Partial Differential Equations in Finance Explained

An Introduction to Computational Finance

palgrave
macmillan

Karel in 't Hout
Department of Mathematics and
Computer Science
University of Antwerp
Antwerp
Belgium

Financial Engineering Explained
ISBN 978-1-137-43568-2   ISBN 978-1-137-43569-9 (eBook)
DOI 10.1057/978-1-137-43569-9

Library of Congress Control Number: 2017934655

Printed on acid-free paper

This Palgrave Macmillan imprint is published by Springer Nature
The registered company is Macmillan Publishers Ltd.
The registered company address is: The Campus, 4 Crinan Street, London, N1 9XW, United Kingdom

Preface

A few years after Black and Scholes [5] derived their famous partial differential equation (PDE) for the fair values of European call and put options, Schwartz [78] considered a finite difference discretization for its approximate solution. Today, the numerical solution of time-dependent PDEs forms one of the pillars of computational finance. Efficient, accurate and stable numerical methods are imperative for financial institutions and companies worldwide. Extensive research is performed, both in academia and industry, into their development, analysis and application. This book is intended as a concise, gentle introduction into this interesting and dynamic field. Its aim is to provide students and practitioners with an easily accessible, practical text explaining main concepts, models, methods and results. The text is organized through a sequence of short chapters. The style is more descriptive than (mathematically) rigorous. Numerous examples and numerical experiments are given to illustrate results. Only some elementary knowledge of mathematics, notably calculus and linear algebra, is assumed.

The numerical solution processes in this book are obtained following the popular method of lines (MOL) approach. Here a given time-dependent PDE is semidiscretized on a grid by finite difference formulas, which yields a large system of ordinary differential equations (ODEs). Subsequently, a suitable temporal discretization method is applied, which defines the full discretization.

Chapters 1 and 2 introduce financial option valuation and partial differential equations. Next, the MOL approach is elaborated in Chapters 3–8. Much attention is paid to studying stability and convergence of the various discretizations. Important special topics, such as boundary conditions, nonuniform grids, the treatment of nonsmooth initial data and approximation of the so-called Greeks, are included in the discussion. In this part the Black–Scholes PDE serves as the prototype equation for the numerical experiments. Examining numerical methods in their application to this equation provides key insight into their properties and performance when applied to many advanced PDEs in contemporary financial mathematics.

After having considered European call and put options as an example, we move on to explore the numerical valuation of more challenging modern types of options: cash-or-nothing options in Chapter 9, barrier options in Chapter 10 and American options in Chapter 11. The latter type of options leads to partial differential inequalities and an additional step in the numerical solution process is required, where so-called linear complementarity problems are solved.

Chapter 12 is devoted to option valuation in the presence of jumps in the underlying asset price evolution. This gives rise to partial integro-differential equations. These equations can be viewed as PDEs with an extra integral term. For their effective numerical solution, operator splitting methods of the implicit-explicit (IMEX) kind are introduced.

Chapter 13 extends the MOL approach to two-dimensional PDEs in finance. Semidiscretization then results in very large systems of ODEs. For the efficient temporal discretization, operator splitting methods of the Alternating Direction Implicit (ADI) kind are discussed. As an example, the numerical valuation of a two-asset option under the Black–Scholes framework is considered.

Most of the chapters conclude with a short section where notes and references to the literature are given. These are intended as pointers to readers who wish to broaden their knowledge or deepen their understanding of the topics under consideration. Supplementary material to this book will be provided on my website.

I am grateful to Peter Forsyth, Sven Foulon, Willem Hundsdorfer, Wim Schoutens, Jari Toivanen and Maarten Wyns for their genuine interest and their valuable suggestions and comments on preliminary versions of this book. Last but not least, I wish to thank Palgrave Macmillan for the pleasant cooperation.

Antwerp, July 2016   Karel in 't Hout

Contents

List of Figures

1

Financial Option Valuation

1.1 Financial Options

A financial option is a so-called derivative product. It is derived from (depends on) a given underlying asset. This underlying asset can be many different things, for example a stock of a company, a commodity or a foreign currency. In precise terms:

> *a financial option is a contract between two parties, the holder and the writer, which gives the holder the right, but not the obligation, to buy from or sell to the writer a given underlying asset at a prescribed price on or before a prescribed time.*

Notice that the holder has the right to exercise, but not the obligation. Hence, the appropriate term "option". The prescribed price in the option contract is called the *strike price* or *exercise price* and shall be denoted by K. The prescribed time in the contract is called the *maturity time* or *expiration time* and shall be denoted by T. By convention, the time of inception of the option is set equal to zero and shall be called *today*.

The above definition encompasses two basic option types: a *call option*, which gives the holder the right to buy the asset, and a *put option*, which gives the holder the right to sell the asset.

© The Author(s) 2017
K. in 't Hout, *Numerical Partial Differential Equations in Finance Explained*,
Financial Engineering Explained, DOI 10.1057/978-1-137-43569-9_1

If exercising by the holder is only allowed at the maturity time, then this is referred to as an *European-style option*. If exercising by the holder is allowed at any given time up to and including the maturity time, then it is called an *American-style option*. This terminology does not have a geographical meaning. In the financial option world, a wide variety of colourful names arises, and more of these will be encountered later on. Unless explicitly stated otherwise, we will always assume that the options under consideration are European-style.

The following example illustrates the natural use of financial options in practice.

Example 1.1.1

Consider company A, working in euros (EUR) and producing certain high-tech machines. Company B, working in US dollars (USD), places an order today with company A for such a machine to be delivered in exactly one year from now for the price of one million USD. Since the actual USD–EUR exchange rate (the value of 1 USD in terms of EUR) in one year's time is unknown, company A faces financial risk in this deal. It can decide to accept this. Alternatively, it can hedge this risk by acquiring, for example from a bank, a put option which gives the right to sell one million USD with maturity time T = 1 year and a certain preferred USD–EUR exchange rate K. Thus company A has locked in a minimal amount of EUR, namely K million, that will be received in one year's time. The value of K is usually chosen close to or equal to today's exchange rate.

The above example shows that options can be employed as an insurance. In general, the use of options can be viewed as redistributing financial risk between parties. The trading of options has grown rapidly over the past decades and is performed both at exchanges and over-the-counter (OTC), that is, directly between two market parties.

It should be clear that options have financial value. The writer of an option is compensated upfront for taking on financial risk. One of the key questions of mathematical finance is:

what, if any, is the fair value of a financial option at inception ?

This question constitutes a fundamental problem, which has been open in the literature for many years. Pioneers of modern option valuation are Bachelier [3], Black and Scholes [5] and Merton [62]. In their celebrated work, Black and Scholes [5] succeeded in answering the above question, under a number of assumptions, for call and put options.

1.2 The Black–Scholes PDE

Let S_τ denote the price of the underlying asset at time $\tau \in [0, T]$. At maturity time the fair value of a call or put option can easily be expressed in terms of the asset price at that time. It is equal to $\phi(S_T)$, where ϕ is the *payoff function* defined by

$$\phi(s) = \begin{cases} \max(s - K, 0) & \text{for } s \geq 0 \quad \text{(call)}, \\ \max(K - s, 0) & \text{for } s \geq 0 \quad \text{(put)}. \end{cases} \tag{1.1}$$

For example, for a put option, if the asset price at maturity S_T is lower than the strike price K, then the holder will exercise this option and sell the asset for K, making a profit of $K - S_T$. On the other hand, if S_T is higher than K, then the holder will not exercise this option, as he/she can obviously sell the asset in the market for a better price than K. In this case the put option is worthless.

Figure 1.1 displays the graphs of the payoff functions for call and put options on the s-domain $[0, 3K]$. Clearly, the two payoffs are piecewise linear functions, both with a kink at the strike K.

The primary objective in mathematical finance is to determine the fair option value today, that is, $\tau = 0$. This value is not obvious, since future asset prices are unknown; compare Example 1.1.1. In order to arrive at a fair option value, Black and Scholes assumed that the asset price evolution is given by a stochastic process, so that S_τ is a random variable for each $\tau \in (0, T]$. More precisely, they considered

$$S_\tau = S_0 \, e^{(\mu - \frac{1}{2}\sigma^2)\tau + \sigma W_\tau} \quad (0 \leq \tau \leq T), \tag{1.2}$$

where W_τ ($\tau \geq 0$) denotes the standard Brownian motion or Wiener process, see Appendix A. The $\sigma > 0$ and μ are real constants, called the *volatility* and the *drift rate*, respectively. The stochastic process (1.2) is referred to as the *geometric Brownian motion*. If S_0 is nonzero, then

$$\ln\left(\frac{S_\tau}{S_0}\right) = (\mu - \tfrac{1}{2}\sigma^2)\tau + \sigma W_\tau.$$

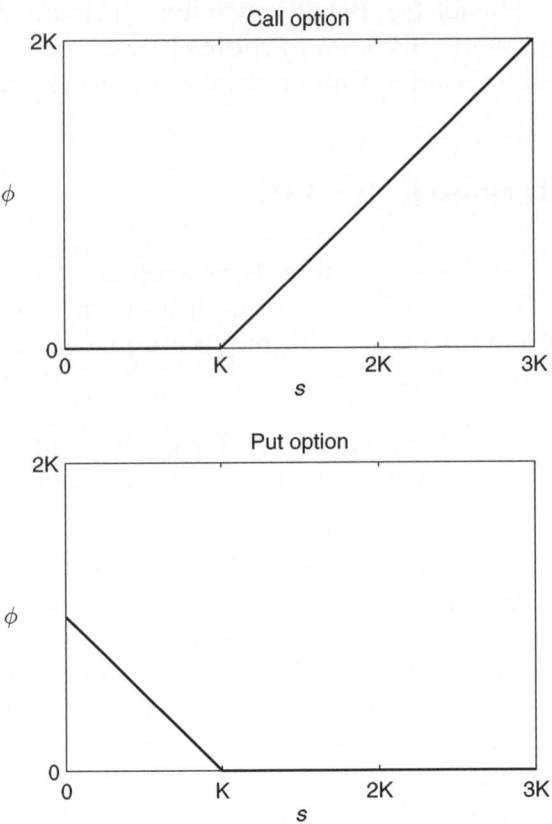

Figure 1.1   Payoff functions for call and put options on $[0, 3K]$

Hence, $\ln(S_\tau/S_0)$ is normally distributed with mean $(\mu - \frac{1}{2}\sigma^2)\tau$ and variance $\sigma^2\tau$. In view of this, the random variable S_τ is said to have a *lognormal distribution*. It can be shown that the expected value of S_τ is given by

$$\mathbb{E}[S_\tau] = S_0 e^{\mu\tau}.$$

Thus the drift μ is the rate of return on the expected future asset prices. Next, the volatility σ forms a measure for the uncertainty in the future asset prices. An increase in the value of σ yields a higher probability for large fluctuations.

In addition to the geometric Brownian motion (1.2), several other assumptions were made by Black and Scholes about the market in which the option and the underlying asset are traded. For our purposes here we mention the *risk-free interest rate r*. This is the theoretical interest rate that holds without any risk of financial loss. It is assumed to be known and constant and so that a cash investment D_0 today grows deterministically in time according to

$$D(\tau) = D_0 e^{r\tau}.$$

A subsequent main assumption in the Black–Scholes framework is *no-arbitrage*. This condition states that there is no possibility of a risk-free return in the market that is greater than that provided by the risk-free rate r. In other words, if one wishes to achieve a greater return than that provided by r, then some risk of financial loss is involved.

Under the Black–Scholes assumptions it can be proved that there exists a unique deterministic real function u of two real variables s and t such that $u(s, t)$ is the fair value at time $\tau = T - t$ of a call or put option if at that time the asset price equals s, and this function satisfies

$$\frac{\partial u}{\partial t}(s, t) = \tfrac{1}{2}\sigma^2 s^2 \frac{\partial^2 u}{\partial s^2}(s, t) + rs\frac{\partial u}{\partial s}(s, t) - ru(s, t) \tag{1.3}$$

for $s > 0$ and $0 < t \leq T$. Here $\partial u/\partial t$ denotes the first-order partial derivative, in the mathematical sense, of u with respect to the independent variable t and $\partial u/\partial s$, $\partial^2 u/\partial s^2$ denote the first- and second-order partial derivatives, respectively, of u with respect to the independent variable s. The simple change of variable from time τ to time-till-maturity t has been done to obtain a formulation that is slightly more convenient.

Equation (1.3) is called the *Black–Scholes partial differential equation (PDE)*. This seminal result was established in [5, 62] in 1973. Modern texts including derivations of (1.3) are for example [47, 66, 80, 90]. In the following we discuss several observations and subsequent results related to the Black–Scholes PDE.

First notice that the function u yields the fair option value for any possible asset price at any given time that the option is in existence. It can be shown that if the market price is not the fair value, then there is an arbitrage opportunity.

As a next observation, the drift rate μ from the asset price process (1.2) does not appear in the Black–Scholes PDE. This is an equally surprising and fundamental result, which has led to the important so-called risk-neutral option valuation theory. It is beyond the scope of this book to discuss this topic, but see for example the books cited above.

In addition to the Black–Scholes PDE, two further conditions hold for the call or put option value function u. The first one is a direct consequence of the fact that the fair option value is known at maturity, that is, if $t = 0$. It is given by the payoff function (1.1) and yields the *initial condition*

$$u(s, 0) = \phi(s) \quad \text{for } s > 0. \tag{1.4}$$

The second one follows from the fact that if the asset price is ever zero, then by the geometric Brownian motion (1.2) it remains zero until maturity. Hence, the fair value at maturity of a call option is equal to zero and of a put option equal to K. By the no-arbitrage assumption, these values are *discounted* with the risk-free interest rate r to any given time $\tau = T - t$, yielding the *boundary condition*

$$u(0, t) = \begin{cases} 0 & \text{for } 0 \le t \le T \quad \text{(call)}, \\ e^{-rt}K & \text{for } 0 \le t \le T \quad \text{(put)}. \end{cases} \tag{1.5}$$

The equations (1.3), (1.4) and (1.5) together form an *initial-boundary value problem for the Black–Scholes PDE* and uniquely determine the option value function u.

Exact solutions to initial-boundary value problems for PDEs are often not at hand in (semi-)closed analytical form. The present case is a useful exception, however. For call and put options, Black and Scholes derived the famous formula

$$u(s, t) = s\,\mathcal{N}(d_1) - e^{-rt}K\mathcal{N}(d_2) \quad \text{(call)}, \tag{1.6a}$$

$$u(s, t) = -s\,\mathcal{N}(-d_1) + e^{-rt}K\mathcal{N}(-d_2) \quad \text{(put)}, \tag{1.6b}$$

for $s > 0$, $0 < t \le T$ with

$$d_1 = \frac{\ln(s/K) + (r + \frac{1}{2}\sigma^2)t}{\sigma\sqrt{t}}, \quad d_2 = d_1 - \sigma\sqrt{t}, \tag{1.6c}$$

where $\mathcal{N}$ denotes the standard normal cumulative distribution function,

$$\mathcal{N}(y) = \frac{1}{\sqrt{2\pi}} \int_{-\infty}^{y} e^{-\frac{1}{2}x^2} dx \quad (y \in \mathbb{R}).$$

There is a neat relation, the *put-call parity*, which links the fair put and call option values on the same underlying for the same strike K and maturity T:

$$put(s, t) + s = call(s, t) + e^{-rt}K \quad (s \geq 0, \ 0 \leq t \leq T). \tag{1.7}$$

This relation can be proved by an elementary argument using the no-arbitrage assumption and holds in a general setting for European-style options.

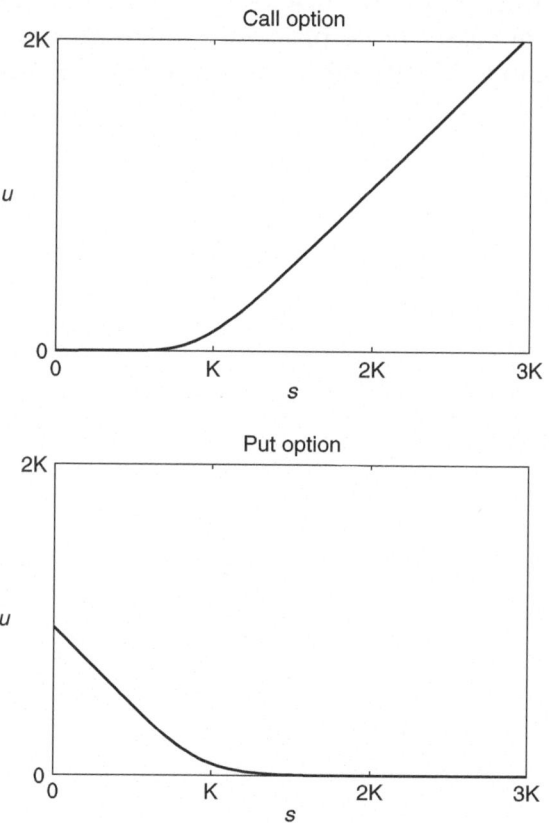

Figure 1.2 Exact call and put option value functions on $[0, 3K]$ for $t = T$ and parameter set (1.8)

As an illustration, Figure 1.2 displays the graphs of the call and put option value functions given by the Black–Scholes formula (1.6) on $[0, 3K]$ for $t = T$ and parameters

$$K = 100, \ T = 1, \ r = 0.05, \ \sigma = 0.25. \tag{1.8}$$

In financial practice there is a big demand for more advanced options that are tailored to the specific needs of investors. These are referred to as *exotic options*. It turns out that, under the Black–Scholes assumptions, the PDE (1.3) is not just fulfilled for the fair values of standard call and put options, but also for a wide range of exotic options. For these only the domain and the initial and boundary conditions for (1.3) change. A number of examples will be seen in the course of this book. If one moves outside of the Black–Scholes framework, for instance when considering asset price processes different from the geometric Brownian motion (1.2), then the option valuation PDE (1.3) itself will change.

2

Partial Differential Equations

2.1 Convection-Diffusion-Reaction Equations

As opposed to ordinary differential equations (ODEs), partial differential equations (PDEs) concern functions of multiple independent variables. In this chapter we consider PDEs of the *time-dependent convection-diffusion-reaction* kind,

$$\frac{\partial u}{\partial t}(s, t) = d(s)\frac{\partial^2 u}{\partial s^2}(s, t) + c(s)\frac{\partial u}{\partial s}(s, t) - r(s)u(s, t) \qquad (2.1)$$

for $S_{min} < s < S_{max}$ and $0 < t \leq T$. Here S_{min}, S_{max} are given real values or $\pm\infty$ and c, d, r denote given real-valued functions where d is always assumed to be nonnegative. The real-valued function u is the unknown.

Most of the PDEs that arise in contemporary financial option valuation theory are of the convection-diffusion-reaction kind. Clearly the choice

$$c(s) = rs, \quad d(s) = \tfrac{1}{2}\sigma^2 s^2, \quad r(s) \equiv r$$

yields the Black–Scholes PDE (1.3). Further, one has $S_{min} = 0$ and $S_{max} = \infty$ for standard call and put options.

In the literature, often an alternative notation by means of subscripts is used for the partial derivatives of a function. This gives a slightly

© The Author(s) 2017
K. in 't Hout, *Numerical Partial Differential Equations in Finance Explained*,
Financial Engineering Explained, DOI 10.1057/978-1-137-43569-9_2

shorter formulation for (2.1), which we will often employ,

$$u_t(s, t) = d(s)u_{ss}(s, t) + c(s)u_s(s, t) - r(s)u(s, t).$$

Time-dependent convection-diffusion-reaction equations are ubiquitous in science and engineering. In applications in physics and chemistry the u_{ss}-part represents a diffusion process, the u_s-part a convection process (for instance, transport due to fluid flow) and the u-part a reaction process. In this context, s is often called the *space variable* and t the *time variable*. Since s lies on the real line, the PDE (2.1) is said to be *one-dimensional*.

In financial mathematics, the convection and diffusion parts correspond to, respectively, the drift and the volatility in the underlying asset price process (under the so-called risk-neutral probability measure). The reaction term represents discounting. The independent variables s and t stand for the underlying asset price and the time-till-maturity. For these two variables it is common practice to use also the physical terminology above.

We always assume that the PDE (2.1) is supplemented with an *initial condition*,

$$u(s, 0) = u_0(s) \quad \text{for } S_{min} < s < S_{max}, \tag{2.2}$$

where u_0 denotes a given function. Thus $u(s, t)$ is prescribed for $t = 0$. In physics and chemistry, the function u_0 gives the initial state of the quantity under consideration. If (2.1) represents a financial option valuation PDE, then u_0 is equal to the payoff function ϕ, that is, the fair option value at maturity.

2.2 The Model Equation

To gain first insight into possible solutions of convection-diffusion-reaction equations, we let the spatial domain $(S_{min}, S_{max}) = \mathbb{R}$ and discuss the *model equation*,

$$u_t(s, t) = du_{ss}(s, t) + cu_s(s, t) - ru(s, t), \tag{2.3}$$

with real constants c, d, r. We consider the convection, diffusion and reaction parts separately, starting with the simplest case.

Reaction equation The pure reaction equation $u_t(s,t) = -ru(s,t)$ is directly solved and yields the exact solution

$$u(s,t) = e^{-rt}u_0(s) \quad \text{for } s \in \mathbb{R}, \ 0 \le t \le T.$$

In a financial context, this represents discounting (with the risk-free rate).

Convection equation The pure convection equation $u_t(s,t) = cu_s(s,t)$ also has a simple exact solution, namely

$$u(s,t) = u_0(s+ct) \quad \text{for } s \in \mathbb{R}, \ 0 \le t \le T.$$

For any given fixed t, the graph of $u(\cdot,t)$ is a shift of the graph of the initial function u_0 with $-ct$ units. If $c > 0$ then the shift is to the left, whereas if $c < 0$ then it is to the right. When s is interpreted as spatial position and t as time, then $-c$ represents velocity.

Diffusion equation The pure diffusion equation $u_t(s,t) = du_{ss}(s,t)$ with $d > 0$ is often called the *heat equation* and does in general not admit closed analytical formulas for its exact solutions. A particular solution is given by

$$p(s,t) = \frac{1}{\sqrt{4\pi dt}} \exp\left(-\frac{s^2}{4dt}\right) \quad \text{for } s \in \mathbb{R}, \ 0 < t \le T.$$

The above is called the *fundamental solution* or *Green's function*. For clarity, dt denotes here the product of d and t. If t tends to zero, then $p(\cdot,t)$ becomes the Dirac delta function. The solution p can be viewed as the probability density function of a normal distribution with mean 0 and variance $2dt$. The general solution to the heat equation can be expressed in semi-closed analytical form,

$$u(s,t) = \int_{-\infty}^{\infty} p(s-x,t)u_0(x)dx. \tag{2.4}$$

Hence, $u(s, t)$ can be regarded as a weighted average of u_0 or as the expected value of a certain random variable. We mention that by using the Green's function it is possible to derive the Black–Scholes formula (1.6), see for example [90].

2.3 Boundary Conditions

In computational practice, to render the numerical solution feasible, the spatial domain $(S_{\min}, S_{\max})$ is taken as bounded and conditions on u are prescribed at the boundary points of this domain.

For various financial options boundedness of the spatial domain intrinsically holds. For instance, it is often fulfilled for barrier options, which shall be discussed in Chapter 10. If the spatial domain is intrinsically unbounded, such as for standard call and put options, then it is common to truncate this domain.

Boundary conditions for PDEs appear in several types. The two best known are the *Dirichlet boundary condition* and the *Neumann boundary condition*. These conditions prescribe, respectively, the values of $u(s, t)$ and the values of the first-order derivative $u_s(s, t)$ at the pertinent spatial boundary.

As an illustration, consider call and put options and a truncated domain $(0, S_{\max})$. At the lower boundary $s = 0$ the Dirichlet condition (1.5) applies. At the upper boundary $s = S_{\max}$ a common Dirichlet condition is specified by

$$u(S_{\max}, t) = \begin{cases} S_{\max} - e^{-rt}K & \text{for } 0 \le t \le T \quad \text{(call)}, \\ 0 & \text{for } 0 \le t \le T \quad \text{(put)}, \end{cases} \tag{2.5}$$

and a Neumann condition by

$$u_s(S_{\max}, t) = \begin{cases} 1 & \text{for } 0 \le t \le T \quad \text{(call)}, \\ 0 & \text{for } 0 \le t \le T \quad \text{(put)}. \end{cases} \tag{2.6}$$

The two conditions (2.5), (2.6) are visually in accordance with the graphs of u in Figure 1.2 whenever $S_{\max}$ is sufficiently large. These conditions only provide approximations, however, to the actual option value function and its first derivative to s. But by taking $S_{\max}$ sufficiently large, the approximation error can be made arbitrarily small.

The conditions (2.5), (2.6) have a natural financial interpretation. Consider for example a put. If the asset price is ever large relative to the strike, then it is very likely to remain large until maturity and a put option will be worthless. This explains the homogeneous (that is, zero) condition in (2.5). Further, a small change in the asset price will have essentially no influence on the put option value, yielding the homogeneous condition in (2.6). These arguments remain valid for many other asset price processes than the geometric Brownian motion. Thus (2.5) and (2.6) are generic in this sense.

For exotic options, deducing accurate Dirichlet or Neumann conditions at the truncated boundaries can be a difficult task. A popular alternative is to impose the *linear boundary condition*,

$$u_{ss}(S_{\max}, t) = 0 \quad \text{for } 0 \le t \le T. \tag{2.7}$$

This is also called the *zero gamma condition* (compare Chapter 6). Clearly, (2.7) states that the second derivative of the option value to s vanishes at $s = S_{\max}$. This represents a linear dependence of the option value on the underlying asset price and holds for a broad variety of financial options whenever $S_{\max}$ is sufficiently large. Figure 1.2 illustrates this for call and put options.

The actual choice of the truncated domain is often done heuristically in practice. It should be large enough such that the approximation errors at the truncated boundaries have a negligible effect on the option values in the *region of interest (ROI)*. The latter is a subdomain of underlying asset prices $S_0 = s$ that is of actual, practical interest. For example, for call and put options, a possible region of interest is $\frac{1}{2}K < s < \frac{3}{2}K$. A common choice for $S_{\max}$ in the Black–Scholes framework is then

$$S_{\max} = se^{(r - \frac{1}{2}\sigma^2)T + \alpha\sigma\sqrt{T}} \tag{2.8}$$

with $s = \frac{3}{2}K$ and $\alpha \approx 3$. This choice is prompted by the fact that, in the risk-neutral setting, the probability that $S_T > S_{\max}$ (with $S_0 = s$) is low; it corresponds to a so-called three-sigma event.

Finally, observe that if s tends to zero in the Black–Scholes PDE (1.3), then the convection and diffusion parts both vanish. Accordingly, the PDE is said to *degenerate* at the boundary $s = 0$. It turns out that degeneracy is a typical feature with financial option valuation PDEs. If $s = 0$, then the Black–Scholes PDE reduces to a pure reaction equation, and its exact solution in the case of call and put options is given by (1.5).

2.4   Notes and References

The PDE (2.1) is linear, that is, if u and v are any given two solutions, then also any linear combination of u and v forms a solution.

Throughout this book we shall assume that the initial-boundary value problems for PDEs under consideration always possess a unique classical solution. Existence and uniqueness results are provided in, for example, the book [22].

In financial mathematics, the fair values of options are often expressed as expected discounted payoff values under the so-called risk-neutral probability measure, see for example the book [80]. It is the Feynman–Kac theorem that provides the mathematical connection between these expectations and solutions to convection-diffusion-reaction equations. This key theorem is stated in for example [59, 68, 80] and has been outlined in Appendix B.

The topic of domain truncation is addressed for example in [54, 92].

3

Spatial Discretization I

3.1 Method of Lines

For the numerical solution of initial-boundary value problems for convection-diffusion-reaction equations (2.1) the *method of lines (MOL)* forms a flexible and versatile approach. It is widely employed in practice and is popular in particular in computational finance. The MOL consists of two general, consecutive steps:

(S) discretization in the space variable s,

(T) discretization in the time variable t.

Step (S) is referred to as *spatial discretization* or *semidiscretization*. In this step the initial-boundary value problem for the PDE is discretized on a finite grid in the s-domain. This leads to an initial value problem for a (large) system of ODEs, the so-called *semidiscrete system*. Step (T) is referred to as *temporal discretization*. Here the semidiscrete system is discretized on a finite grid in the t-domain and defines the actual, fully discrete approximations on the obtained Cartesian grid in the (s, t)-domain. A sample (s, t)-grid is shown in Figure 3.1.

The present and the subsequent two chapters deal with step (S). We shall discuss several basic semidiscretizations of initial-boundary value problems for PDEs (2.1). In this chapter we commence with the

© The Author(s) 2017   **15**
K. in 't Hout, *Numerical Partial Differential Equations in Finance Explained*,
Financial Engineering Explained, DOI 10.1057/978-1-137-43569-9_3

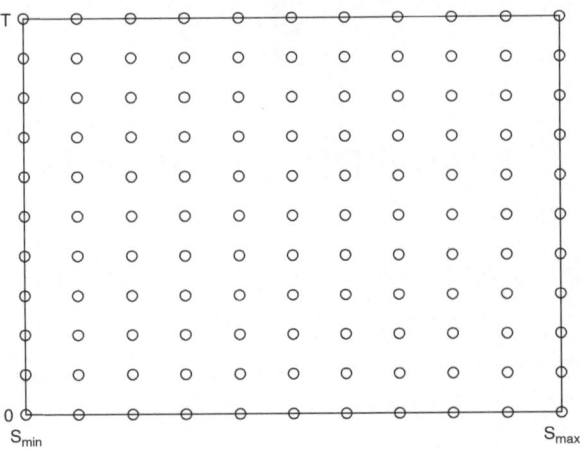

Figure 3.1   Sample grid in the (s, t)-domain, indicated by circles

model equation (2.3). Along with the initial condition (2.2), a so-called *periodic (boundary) condition* is taken here for simplicity,

$$u(s + 1, t) = u(s, t) \quad \text{for all } s \in \mathbb{R}, \ 0 \le t \le T. \tag{3.1}$$

Thus the solution is assumed to be periodic in the spatial variable with period 1. The model equation with periodic condition forms a natural starting point in the numerical literature as it enables a rigorous theoretical stability analysis that provides important practical insight. Notice that (3.1) is not a boundary condition in the strict sense, but it is still commonly named as such. Actual boundary conditions will be discussed in the next chapter.

In view of the periodicity, it suffices to consider semidiscretization on the spatial interval $(0, 1]$. Let $m \ge 3$ be any given integer, let the *spatial mesh width* $h = 1/m$ and let *spatial grid points* $s_i = ih$ for $i = 0, 1, 2, \ldots, m$. The spatial discretizations under consideration in this book are based upon *finite difference formulas*. They yield approximations $U_i(t)$ to $u(s_i, t)$ for $1 \le i \le m$, $0 < t \le T$. By the initial condition (2.2), the values at $t = 0$ are directly known,

$$U_i(0) = u_0(s_i) \quad \text{for } 1 \le i \le m. \tag{3.2}$$

3.2 Finite Difference Formulas

In the following we formulate several basic semidiscretizations for the model convection and diffusion equations separately. These are subsequently combined so as to arrive at semidiscretizations for the full model equation (2.3).

By classical real analysis, the first derivative f' of any given smooth function $f : \mathbb{R} \to \mathbb{R}$ at any point $s \in \mathbb{R}$ is approximated by

$$f'(s) \approx \frac{f(s) - f(s - h)}{h} \qquad (3.3)$$

whenever $h > 0$ is small. The right-hand side of (3.3) is a finite difference quotient, involving two values of f. It is called the *first-order backward formula*. This formula can be interpreted as the slope of the line segment between the points $(s - h, f(s - h))$ and $(s, f(s))$ on the graph of f, see Figure 3.2.

Consider now the pure model convection equation $u_t(s, t) = c u_s(s, t)$ with periodic condition. Applying (3.3) to the spatial derivative term in this equation at grid point $s = s_i$, yields the approximate relation

$$u_t(s_i, t) \approx c \, \frac{u(s_i, t) - u(s_{i-1}, t)}{h} \qquad (1 \le i \le m, \ 0 < t \le T).$$

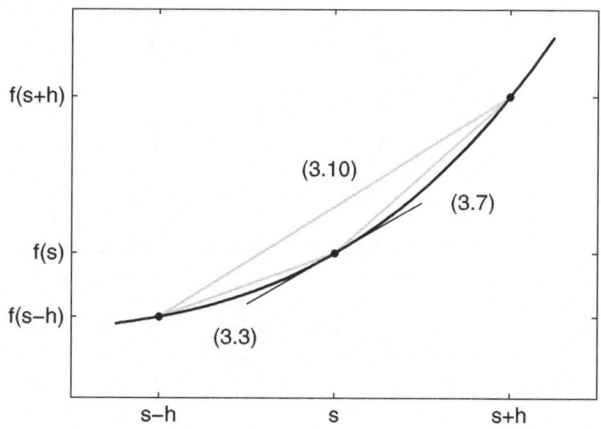

Figure 3.2   Geometric interpretation of the finite difference formulas (3.3), (3.7), (3.10) for the first derivative $f'(s)$

It is then a natural idea to *define* approximations $U_i(t)$ to $u(s_i, t)$ through the exact relation

$$U_i'(t) = c \frac{U_i(t) - U_{i-1}(t)}{h} \quad (1 \leq i \leq m, \ 0 < t \leq T), \qquad (3.4)$$

with $U_0(t) = U_m(t)$ by periodicity. This constitutes a first semidiscretization of the model convection equation with periodic condition. Observe that (3.4) involves only one continuous independent variable, namely t. Together with (3.2), one has obtained an initial value problem for a system of ODEs. It is convenient to consider (3.4) in vector form. Let

$$U(t) = (U_1(t), U_2(t), \ldots, U_m(t))^{\mathrm{T}} \quad \text{and} \quad U'(t) = (U_1'(t), U_2'(t), \ldots, U_m'(t))^{\mathrm{T}},$$

where the symbol ^{T} denotes taking the transpose. Then the semidiscrete system (3.4) can be formulated as

$$U'(t) = AU(t) \quad (0 < t \leq T), \qquad (3.5)$$

with $m \times m$ matrix A given by

$$A = \frac{c}{h} \begin{pmatrix} 1 & & & & -1 \\ -1 & 1 & & & \\ & \ddots & \ddots & & \\ & & -1 & 1 & \\ & & & -1 & 1 \end{pmatrix}. \qquad (3.6)$$

Instead of (3.3), the first derivative of f can also be approximated by

$$f'(s) \approx \frac{f(s+h) - f(s)}{h}. \qquad (3.7)$$

The right-hand side of (3.7) is called the *first-order forward formula* and can be interpreted as the slope of the line segment between the points $(s, f(s))$ and $(s + h, f(s + h))$, see Figure 3.2. It gives rise to the semidiscrete system

$$U_i'(t) = c\,\frac{U_{i+1}(t) - U_i(t)}{h} \quad (1 \le i \le m,\; 0 < t \le T), \tag{3.8}$$

with $U_{m+1}(t) = U_1(t)$ by periodicity. This can be written in the form (3.5) with

$$A = \frac{c}{h}\begin{pmatrix} -1 & 1 & & & \\ & -1 & 1 & & \\ & & \ddots & \ddots & \\ & & & -1 & 1 \\ 1 & & & & -1 \end{pmatrix}. \tag{3.9}$$

For reasons of stability, which shall be explained in the next section, the first-order backward formula should only be used if $c < 0$ and the first-order forward formula only if $c > 0$. The combination is named the *first-order upwind formula*.

Considering values of f for arguments on both sides of s, yields the approximation

$$f'(s) \approx \frac{f(s + h) - f(s - h)}{2h}. \tag{3.10}$$

The right-hand side of (3.10) is called the *second-order central for-mula (for convection)* and is also visualized in Figure 3.2. It leads to the semidiscrete system

$$U_i'(t) = c\,\frac{U_{i+1}(t) - U_{i-1}(t)}{2h} \quad (1 \le i \le m,\; 0 < t \le T), \tag{3.11}$$

with $U_0(t) = U_m(t)$ and $U_{m+1}(t) = U_1(t)$, and is of the form (3.5) with

$$A = \frac{c}{2h}\begin{pmatrix} 0 & 1 & & & -1 \\ -1 & 0 & 1 & & \\ & \ddots & \ddots & \ddots & \\ & & -1 & 0 & 1 \\ 1 & & & -1 & 0 \end{pmatrix}. \tag{3.12}$$

The *truncation error* of any given finite difference formula is defined as the exact derivative value minus the pertinent finite difference quotient. As their names suggest, the truncation errors for the first-order formulas are $\mathcal{O}(h)$, whereas for the second-order central formula it is $\mathcal{O}(h^2)$. This is readily verified by using Taylor's theorem. Accordingly, faster convergence to the PDE solution u as $h \downarrow 0$ can be expected when applying the latter formula.

For the second derivative of any given smooth function $f : \mathbb{R} \to \mathbb{R}$ one has the approximation

$$f''(s) \approx \frac{f(s-h) - 2f(s) + f(s+h)}{h^2}. \tag{3.13}$$

The right-hand side of (3.13) is called the *second-order central formula (for diffusion)*. Considering this formula in the case of the pure model diffusion equation $u_t(s,t) = d u_{ss}(s,t)$ with periodic condition leads to the semidiscrete system

$$U_i'(t) = d \frac{U_{i-1}(t) - 2U_i(t) + U_{i+1}(t)}{h^2} \quad (1 \le i \le m, \ 0 < t \le T), \tag{3.14}$$

where $U_0(t) = U_m(t)$ and $U_{m+1}(t) = U_1(t)$. This is again of the form (3.5), with

$$A = \frac{d}{h^2} \begin{pmatrix} -2 & 1 & & & 1 \\ 1 & -2 & 1 & & \\ & \ddots & \ddots & \ddots & \\ & & 1 & -2 & 1 \\ 1 & & & 1 & -2 \end{pmatrix}. \tag{3.15}$$

For finite difference discretization of diffusion, the second-order central formula is mostly employed in practice. The truncation error for this formula is $\mathcal{O}(h^2)$.

To obtain semidiscretizations of the full model equation (2.3), one can just combine semidiscretizations of its individual terms. If A_1 and A_2 are any given matrices corresponding to the semidiscretized convection and diffusion terms, respectively, then a semidiscretization of (2.3) is given by $A = A_2 + A_1 - rI$ with I the $m \times m$ identity matrix, where the $-rI$ part stems from the reaction term.

3.3 Stability

In this section we study the stability of the semidiscrete systems constructed in Section 3.2. All these systems are of the form (3.5). By virtue of the periodicity condition, all pertinent matrices A are *normal*, that is, $AA^T = A^TA$. Consequently, for each matrix A, it holds that $A = V\Lambda V^{-1}$ with certain unitary matrix V and diagonal matrix $\Lambda = \text{diag}(\lambda_1, \lambda_2, \ldots, \lambda_m)$. The columns of V are the eigenvectors of A and the diagonal entries of Λ are the eigenvalues of A.

Let $\mathbf{i}$ denote the imaginary unit. For the four matrices A from Section 3.2 explicit expressions for their eigenvalues λ_k $(1 \leq k \leq m)$ are known:

$$(3.6) : \lambda_k = \frac{c}{h}\left(1 - \cos\left(2\pi kh\right) + \mathbf{i}\sin\left(2\pi kh\right)\right),$$

$$(3.9) : \lambda_k = \frac{c}{h}\left(\cos\left(2\pi kh\right) - 1 + \mathbf{i}\sin\left(2\pi kh\right)\right),$$

$$(3.12) : \lambda_k = \mathbf{i}\frac{c}{h}\sin\left(2\pi kh\right),$$

$$(3.15) : \lambda_k = -\frac{4d}{h^2}\sin^2\left(\pi kh\right).$$

Hence, for the first-order backward and forward formulas all eigenvalues lie on a circle in the complex plane with radius $\frac{|c|}{h}$ and midpoints $(\frac{c}{h}, 0)$ and $(-\frac{c}{h}, 0)$, respectively. For the second-order central formula for convection all eigenvalues lie on the imaginary axis, and for the second-order central formula for diffusion all eigenvalues lie on the negative real axis.

The stability analysis of semidiscrete systems deals with the propagation forward in time of perturbations in the initial vector. A given semidiscretization is said to be *stable* if the error incurred at any given time $t > 0$ can be bounded by a moderate constant multiplied by the error at the initial time $t = 0$, where the constant is independent of the spatial mesh width h and the initial error.

To measure the sizes of vectors, consider the naturally scaled Euclidean norm

$$\|x\|_2 = \left\{h\sum_{i=1}^{m}|x_i|^2\right\}^{\frac{1}{2}} \quad \text{whenever } x = (x_1, x_2, \ldots, x_m)^T \in \mathbb{C}^m. \quad (3.16)$$

Let U_0, $\widetilde{U}_0$ be any given two initial vectors and let the functions U, $\widetilde{U}$ denote the corresponding solutions to (3.5) with $A = V \Lambda V^{-1}$ as above. By linearity, the difference $\widetilde{U} - U$ is itself solution to the semidiscrete system, with initial vector $\widetilde{U}_0 - U_0$. Consider the transformation

$$Y(t) = V^{-1}(\widetilde{U}(t) - U(t)).$$

Then $\|Y(t)\|_2 = \|\widetilde{U}(t) - U(t)\|_2$ since V is unitary, and there holds

$$Y'(t) = \Lambda Y(t) \quad (0 < t \leq T).$$

The latter ODE system is decoupled,

$$Y_k'(t) = \lambda_k Y_k(t) \quad (0 < t \leq T, 1 \leq k \leq m),$$

with solution

$$Y_k(t) = e^{\lambda_k t} Y_k(0) \quad (0 < t \leq T, 1 \leq k \leq m).$$

It readily follows that

$$\|\widetilde{U}(t) - U(t)\|_2 \leq \max_{1 \leq k \leq m} e^{\Re \lambda_k t} \|\widetilde{U}_0 - U_0\|_2 \quad (0 \leq t \leq T), \qquad (3.17)$$

where $\Re \lambda$ denotes the real part of any complex number λ. The factor in the upper bound (3.17) is sharp when arbitrary initial perturbations $\widetilde{U}_0 - U_0$ are considered. Combining this bound with the eigenvalues of the matrices (3.6), (3.9), (3.12), (3.15) specified above, it is directly seen that the semidiscretization by:

- The first-order backward formula is stable if $c < 0$,
- The first-order forward formula is stable if $c > 0$,
- The second-order central formula for convection is stable for all c,
- The second-order central formula for diffusion is stable for all $d > 0$.

In all four cases the eigenvalues lie in the left-half of the complex plane, so that

$$\|\widetilde{U}(t) - U(t)\|_2 \leq \|\widetilde{U}_0 - U_0\|_2 \quad (0 \leq t \leq T). \qquad (3.18)$$

This stability result, where initial errors do not grow, is often called *contractivity*.

The semidiscretization given by the first-order backward (forward) formula is not stable if $c > 0$ ($c < 0$). This is a consequence of the sharpness of the bound (3.17) and the fact that $\max_{1 \leq k \leq m} \Re \lambda_k \to \infty$ as $h \downarrow 0$.

It can be shown that each of the four stable semidiscretizations converges, as $h \downarrow 0$, to the exact solution u of the corresponding model equation with periodic condition if the initial function u_0 is sufficiently smooth. The convergence behaviour is $\mathcal{O}(h^p)$ in $\| \cdot \|_2$ where p is the order of the finite difference formula.

3.4   Notes and References

The literature on finite difference formulas for partial differential equations is vast. A selection of general texts is [49, 65, 74, 81, 83, 86] and books dealing with financial option valuation applications include [11, 21, 34, 79, 85, 90, 94].

In this book we apply mostly the second-order central formulas for convection and diffusion. There exist however (central and upwind) finite difference formulas of arbitrarily high order that yield stable semidiscretizations; compare the above references.

The stability analysis in this chapter relies upon the matrices A being normal and is related to the famous von Neumann (Fourier) analysis for full discretizations, see for instance [49, 67, 83]. In actual applications, the matrices A are usually nonnormal and then a rigorous, useful stability analysis is often much more involved, see for example [82, 87].

4

Spatial Discretization II

In this chapter we extend the basic semidiscretizations introduced in Chapter 3 to the general convection-diffusion-reaction equation combined with the various boundary conditions from Chapter 2. We then discuss nonuniform spatial grids and consider the numerical treatment of nonsmooth initial functions, which are omnipresent in financial applications. The chapter concludes with a useful mixed central/upwind discretization.

4.1 Boundary Conditions

Let the spatial domain $(S_{\min}, S_{\max})$ be bounded. For ease of presentation it will always be assumed that the general convection-diffusion-reaction equation (2.1) is provided with a Dirichlet condition at the lower boundary $s = S_{\min}$,

$$u(S_{\min}, t) = a_0(t) \quad (0 \le t \le T),$$

where a_0 is a given function.

Let $m \ge 3$ be any given integer and let a spatial mesh width and spatial grid points be given by

$$h = \frac{S_{\max} - S_{\min}}{m} \quad \text{and} \quad s_i = S_{\min} + ih \quad (i = 0, 1, 2, \ldots, m).$$

© The Author(s) 2017
K. in 't Hout, *Numerical Partial Differential Equations in Finance Explained*,
Financial Engineering Explained, DOI 10.1057/978-1-137-43569-9_4

Similarly to Chapter 3, approximations $U_i(t)$ to $u(s_i, t)$ can be defined through finite difference discretization of the convection and diffusion terms in (2.1). The semidiscretizations (3.4), (3.8), (3.11) and (3.14) for the model convection and diffusion equations are directly generalized to variable coefficients upon replacing the convection and diffusion constants c and d in there by $c(s_i)$ and $d(s_i)$, respectively. Next, by the Dirichlet condition at the lower boundary, one immediately has $U_0(t) = a_0(t)$. To complete the semidiscretization it remains to define $U_m(t)$ at the upper boundary $s = S_{\max}$. In the following we successively discuss the three types of boundary conditions from Chapter 2 (i) Dirichlet, (ii) Neumann and (iii) linear. The obtained semidiscrete systems are of the form

$$U'(t) = AU(t) + g(t) \quad (0 < t \leq T), \tag{4.1}$$

where A is a given $v \times v$ matrix, $g(t)$ for each t is a given v-vector, and integer

$$v \in \{m - 1, m\}.$$

For convenience, write $c_i = c(s_i)$, $d_i = d(s_i)$, $r_i = r(s_i)$ for all i.

(i) A *Dirichlet condition* reads

$$u(S_{\max}, t) = a_1(t) \quad (0 \leq t \leq T),$$

with given function a_1. In this case one directly has $U_m(t) = a_1(t)$. Considering for example the second-order central formulas for convection and diffusion at the grid points s_i for $i = 1, 2, \ldots, v$ with $v = m - 1$, a semidiscrete system (4.1) is obtained with tridiagonal matrix A and vector $g(t)$ given by

$$A = \begin{pmatrix} \alpha_1 & \gamma_1 & & & \\ \beta_2 & \alpha_2 & \gamma_2 & & \\ & \ddots & \ddots & \ddots & \\ & & \beta_{v-1} & \alpha_{v-1} & \gamma_{v-1} \\ & & & \beta_v & \alpha_v \end{pmatrix}, \quad g(t) = \begin{pmatrix} \delta_1(t) \\ 0 \\ \vdots \\ 0 \\ \delta_v(t) \end{pmatrix}, \tag{4.2}$$

where

$$\alpha_i = -\frac{2d_i}{h^2} - r_i, \ \beta_i = \frac{d_i}{h^2} - \frac{c_i}{2h}, \ \gamma_i = \frac{d_i}{h^2} + \frac{c_i}{2h} \quad (1 \leq i \leq m-1), \quad (4.3)$$

and

$$\delta_1(t) = \beta_1 a_0(t), \ \delta_{m-1}(t) = \gamma_{m-1} a_1(t).$$

(ii) A *Neumann condition* reads

$$u_s(S_{\max}, t) = b_1(t) \quad (0 \leq t \leq T),$$

with given function b_1. In this case the value of the exact solution u at $s = S_{\max}$ is unknown. Accordingly, an approximation $U_m(t)$ needs to be defined at this grid point. In view of the Neumann condition the convection term $u_s(s, t)$ does not require approximation, since it is directly given there. It therefore suffices to consider the diffusion term. The second-order central formula at $s = s_m$ yields

$$u_{ss}(s_m, t) \approx \frac{u(s_{m-1}, t) - 2u(s_m, t) + u(s_{m+1}, t)}{h^2},$$

with $s_{m+1} = s_m + h$. The latter point lies outside the spatial domain, however, and the question arises how to deal with it. Using the Neumann condition, a useful idea is to replace the value of u at this virtual point by the approximation

$$u(s_{m+1}, t) \approx u(s_{m-1}, t) + 2hb_1(t).$$

This can be regarded as a *linear extrapolation* formula. Inserting into the above, yields

$$u_{ss}(s_m, t) \approx \frac{2u(s_{m-1}, t) - 2u(s_m, t) + 2hb_1(t)}{h^2},$$

where the right-hand side now involves values of u only inside the spatial domain. The following natural definition for the approximation $U_m(t)$ is then obtained,

$$U'_m(t) = 2d_m \frac{U_{m-1}(t) - U_m(t) + hb_1(t)}{h^2} + c_m b_1(t) - r_m U_m(t) \quad (0 < t \leq T).$$

With the second-order central formulas for convection and diffusion at the grid points $s_1, s_2, \ldots, s_{m-1}$ one arrives at a semidiscrete system (4.1) with A and $g(t)$ of the type (4.2) and $\nu = m$. The entries α_i, β_i, γ_i for $1 \leq i \leq m-1$ are defined by (4.3) and the last row of A is specified by

$$\alpha_m = -\frac{2d_m}{h^2} - r_m, \quad \beta_m = \frac{2d_m}{h^2}.$$

Next, the vector $g(t)$ is given by

$$\delta_1(t) = \beta_1 a_0(t), \quad \delta_m(t) = \left(\frac{2d_m}{h} + c_m\right) b_1(t).$$

(iii) The *linear boundary condition* reads

$$u_{ss}(S_{\max}, t) = 0 \quad (0 \leq t \leq T).$$

In this case the value of the exact solution u at the upper boundary is again unknown and an approximation $U_m(t)$ needs to be defined at this grid point. Clearly, the diffusion term vanishes and it remains to consider the convection term. For this term it is common to select the first-order backward formula at $s = s_m$. Thus only points inside the spatial domain are used, as opposed to the second-order central formula. This leads to the following definition for $U_m(t)$,

$$U_m'(t) = c_m \frac{U_m(t) - U_{m-1}(t)}{h} - r_m U_m(t) \quad (0 < t \leq T).$$

A semidiscrete system (4.1) is obtained where A and $g(t)$ are of the type (4.2) with $\nu = m$. The last row of the matrix A is now specified by

$$\alpha_m = \frac{c_m}{h} - r_m, \quad \beta_m = -\frac{c_m}{h},$$

and the vector $g(t)$ is given by

$$\delta_1(t) = \beta_1 a_0(t), \quad \delta_m(t) = 0.$$

4.2 Nonuniform Grids

In financial applications it is often natural and beneficial to employ nonuniform spatial grids instead of uniform grids. In this section we consider how to define suitable nonuniform grids and generalize the finite difference formulas introduced in Chapter 3. Ample numerical illustrations shall be presented in the subsequent chapters.

Nonuniform spatial grids are often used to concentrate grid points near one or more given points of interest in the spatial domain, for instance the strike. Such grids can be conveniently constructed through a given continuous function $\varphi : [\xi_{min}, \xi_{max}] \to [S_{min}, S_{max}]$ with $\varphi(\xi_{min}) = S_{min}$ and $\varphi(\xi_{max}) = S_{max}$ that is strictly increasing and has a relatively gentle slope near the preimages of the points of interest. One then chooses a uniform grid in the artificial ξ-domain and maps this by φ to a nonuniform grid in the actual s-domain,

$$s_i = \varphi(\xi_i) \text{ with } \xi_i = \xi_{min} + i\Delta\xi, \ \Delta\xi = \frac{\xi_{max} - \xi_{min}}{m} \quad (i = 0, 1, 2, \ldots, m).$$

Let $h_i = s_i - s_{i-1}$ for $1 \le i \le m$ denote the variable spatial mesh widths. We always assume that the spatial grid is *smooth* in the sense that there exist real numbers $C_0, C_1, C_2 > 0$ independent of i and m such that the mesh widths satisfy

$$C_0 \Delta\xi \le h_i \le C_1 \Delta\xi \text{ and } |h_{i+1} - h_i| \le C_2 (\Delta\xi)^2.$$

This means that the mesh widths h_i tend to zero at the rate of $\Delta\xi$ and vary gradually. A simple analysis shows that the nonuniform grid is smooth under weak conditions on the mapping φ.

For the numerical experiments in this book we shall consider a particular choice for φ.

Example 4.2.1

Let K be any given point of interest in the spatial domain. Let the mapping φ be defined, via the hyperbolic sine function, by

$$\varphi(\xi) = K + L \sinh(\xi) \quad (\xi_{min} \le \xi \le \xi_{max}),$$

with parameter $L > 0$ and

$$\xi_{min} = \sinh^{-1}((S_{min} - K)/L) \text{ and } \xi_{max} = \sinh^{-1}((S_{max} - K)/L).$$

It is readily verified that φ is continuous and strictly increasing and that the spatial grid generated by this mapping is smooth. The parameter L controls the fraction of grid points s_i that lie in a neighbourhood of K, where a smaller value L yields a denser grid around K. Here we heuristically select $L = K/3$. As an illustration, Figure 4.1 displays the graph of φ if $K = 100$, $S_{min} = 0$, $S_{max} = 3K$. Clearly, the mapping φ has a relatively gentle slope near $\xi = 0$, the preimage of $s = K$. A uniform grid in ξ thus yields a nonuniform grid in s with relatively many points s_i near K.

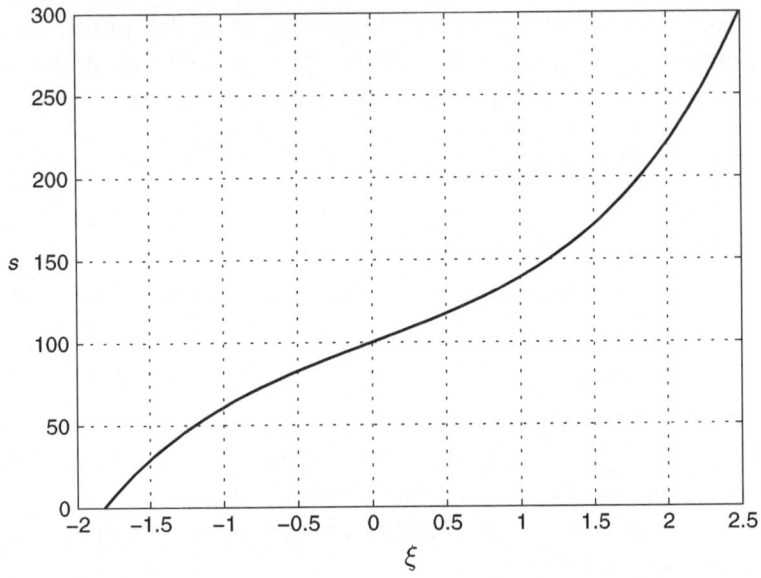

Figure 4.1   Mapping φ defined in Example 4.2.1

Consider next any given smooth function $f : [S_{min}, S_{max}] \to \mathbb{R}$. Then for the first derivative one has the following finite difference approximations pertinent to a nonuniform s-grid.

First-order backward formula:

$$f'(s_i) \approx \frac{f(s_i) - f(s_{i-1})}{h_i} \quad (0 < i \leq m). \tag{4.4}$$

First-order forward formula:

$$f'(s_i) \approx \frac{f(s_{i+1}) - f(s_i)}{h_{i+1}} \quad (0 \leq i < m). \tag{4.5}$$

Second-order central formula A:

$$f'(s_i) \approx \frac{f(s_{i+1}) - f(s_{i-1})}{h_i + h_{i+1}} \quad (0 < i < m). \tag{4.6}$$

Second-order central formula B:

$$f'(s_i) \approx \omega_{i,-1} f(s_{i-1}) + \omega_{i,0} f(s_i) + \omega_{i,1} f(s_{i+1}) \quad (0 < i < m), \tag{4.7}$$

with

$$\omega_{i,-1} = \frac{-h_{i+1}}{h_i(h_i + h_{i+1})} \,, \quad \omega_{i,0} = \frac{h_{i+1} - h_i}{h_i h_{i+1}} \,, \quad \omega_{i,1} = \frac{h_i}{h_{i+1}(h_i + h_{i+1})} \,.$$

For the second derivative one has

Second-order central formula:

$$f''(s_i) \approx \omega_{i,-1} f(s_{i-1}) + \omega_{i,0} f(s_i) + \omega_{i,1} f(s_{i+1}) \quad (0 < i < m), \tag{4.8}$$

with

$$\omega_{i,-1} = \frac{2}{h_i(h_i + h_{i+1})} \,, \quad \omega_{i,0} = \frac{-2}{h_i h_{i+1}} \,, \quad \omega_{i,1} = \frac{2}{h_{i+1}(h_i + h_{i+1})} \,.$$

These finite difference formulas generalize those from Chapter 3 to arbitrary, nonuniform grids. Notice that there are two generalizations, (4.6) and (4.7), of the second-order central formula (3.10) for convection. They are both often employed in computational finance.

The formulas (4.6), (4.7), (4.8) all possess a second-order truncation error for smooth spatial grids. If the grid is not smooth, then the central formulas (4.6) and (4.8) for convection and diffusion, respectively, reduce to only first-order in general. As mentioned previously, however, the latter situation will not be considered in this book.

The actual application of the finite difference formulas (4.4)–(4.8) for the spatial discretization on nonuniform grids of the general convection-diffusion-reaction equation (2.1) is straightforward. Also the numerical treatment of the three types of boundary conditions discussed in Section 4.1 is directly extended to nonuniform grids. In all cases, this leads to a semidiscrete system (4.1) with given matrix A and vector $g(t)$ of the type (4.2).

4.3 Nonsmooth Initial Data

A main feature of financial options is that their payoff functions are usually not smooth. For call and put options, they are continuous on their domain, but not differentiable at the strike. For other types of options the payoffs may not even be continuous at one or more given points. An example is provided by a *cash-or-nothing call option*, which has the payoff

$$\phi(s) = \begin{cases} 0 & \text{for } s < K, \\ D & \text{for } s > K. \end{cases} \tag{4.9}$$

This option pays out a prescribed fixed cash amount $D > 0$ whenever the underlying asset price ends up above the strike price at maturity, $S_T > K$, and it pays out nothing whenever it finishes below the strike price, $S_T < K$.

The nonsmoothness of payoff functions requires careful attention in the numerical solution of initial-boundary value problems for option valuation PDEs (recall that the payoff defines the initial condition). Finite difference approximations rely upon sufficient smoothness of the pertinent functions and nonsmooth payoffs can therefore give rise to an undesirable convergence behaviour of the numerically obtained option prices; this will be illustrated by numerical experiments in Chapter 5. The points where the payoffs are nonsmooth often lie in regions of interest in applications, where one wishes to obtain reliable option prices. Accordingly, it is important to consider an effective numerical treatment.

It can be argued that the pointwise representation of a payoff function ϕ on the spatial grid, which defines the initial vector $U(0)$ for the semidiscrete system, does not capture sufficient information in general about ϕ near the points of nonsmoothness. A useful idea is to replace, for each grid point s_i nearest to any given point of nonsmoothness, the value $\phi(s_i)$ occurring in $U(0)$ by the average value of ϕ over a neighbourhood of s_i. This can be expressed as an integral,

$$\frac{1}{h_{i+1/2}} \int_{s_{i-1/2}}^{s_{i+1/2}} \phi(s)\,ds, \tag{4.10}$$

where

$$s_{i-1/2} = \tfrac{1}{2}(s_{i-1} + s_i), \quad s_{i+1/2} = \tfrac{1}{2}(s_i + s_{i+1}), \quad h_{i+1/2} = s_{i+1/2} - s_{i-1/2}.$$

The value (4.10) is called the *cell average* of ϕ over $[s_{i-1/2}, s_{i+1/2}]$ and the approach is referred to as a *smoothing technique*. In subsequent chapters the effectiveness of this simple technique is demonstrated by ample numerical experiments. For many payoffs, the integral (4.10) is readily calculated. Otherwise a numerical integration method, for instance the trapezoidal rule, can be applied to accurately approximate it.

4.4 Mixed Central/Upwind Discretization

A useful variant of the standard second-order central discretization of the Black–Scholes PDE, with Dirichlet or Neumann boundary conditions and $r > 0$, is given by switching (upfront) from the second-order central formula for convection to the first-order forward formula at each grid point s_i with $1 \leq i \leq \nu$ such that β_i (see Section 4.1) is strictly negative. For a general nonuniform grid, this means that one switches if

$$\frac{s_i}{h_i} < \frac{r}{\sigma^2} \text{ (for central formula A) and } \frac{s_i}{h_{i+1}} < \frac{r}{\sigma^2} \text{ (for central formula B).}$$

Both inequalities can be rewritten in terms of the so-called *cell Péclet number*

$$\frac{c(s)h}{d(s)} \quad \text{with} \quad c(s) = rs, \ d(s) = \tfrac{1}{2}\sigma^2 s^2.$$

If an inequality holds, one says that the Péclet condition is violated. For typical values of r, σ this happens for at most a small fraction of the set of grid points. Moreover, these points usually lie far away from the region of interest. Therefore, little adverse impact on the spatial accuracy in this region is to be expected. The *mixed central/upwind* discretization defined above possesses the important feature that $-A$ is an M-matrix. The slightly stronger condition is also fulfilled that

$$\begin{cases} \beta_i \geq 0 & (2 \leq i \leq v), \\ \gamma_i \geq 0 & (1 \leq i \leq v-1), \\ \alpha_1 + \gamma_1 \leq -r, \\ \alpha_i + \beta_i + \gamma_i = -r & (2 \leq i \leq v-1), \\ \alpha_v + \beta_v \leq -r. \end{cases} \tag{4.11}$$

This condition implies many favourable properties for the semidiscretization. In particular, contractivity in the maximum norm $\|\cdot\|_\infty$ holds: if U_0, $\widetilde{U}_0$ are any given two initial vectors and U, $\widetilde{U}$ are the corresponding solutions to (4.1), then

$$\|\widetilde{U}(t) - U(t)\|_\infty \leq \|\widetilde{U}_0 - U_0\|_\infty \quad (0 \leq t \leq T). \tag{4.12}$$

Furthermore, the semidiscretization is *positivity preserving*:[1]

$$U_0 \geq 0 \text{ and } g(t) \geq 0 \ (0 \leq t \leq T) \quad \Longrightarrow \quad U(t) \geq 0 \ (0 \leq t \leq T), \tag{4.13}$$

where inequalities for vectors are to be interpreted componentwise. The semidiscretization is also free from spurious oscillations. We shall return to these properties for the full discretization in Chapter 7. Even though it has been considered here for the Black–Scholes PDE, the mixed central/upwind spatial discretization is readily extended and beneficial in many, more advanced applications.

---

[1] More precisely phrased: nonnegativity preserving.

4.5   Notes and References

For a subtle analysis concerning the discretization of the linear boundary condition at the upper boundary for the Black–Scholes PDE, we refer to [40, 92].

The type of nonuniform grid constructed in Example 4.2.1 is frequently used in financial applications, see for instance [85].

The terminology of formula A and formula B for (4.6) and (4.7), respectively, has been adopted from [88].

The treatment of nonsmooth initial data in the semidiscretization of financial PDEs has been discussed for instance in [71, 85]. A general, classical reference is for example [55].

The mixed central/upwind discretization has been considered in for example [33, 40, 70, 92]. Proofs of the properties (4.12), (4.13) under (4.11) can be found in for instance [49]. A general reference to M-matrices is [4].

5

Numerical Study: Space

In this chapter we study by numerical experiments the performance of spatial discretizations introduced in Chapters 3 and 4. Here a call option under the Black–Scholes framework, discussed in Chapter 1, is considered. This forms a prototype for many, more advanced financial applications and the obtained insights are of general importance.

Consider the Black–Scholes PDE (1.3) on the truncated spatial domain

$$(S_{\min}, S_{\max}) = (0, 3K)$$

and choose Dirichlet conditions (1.5) and (2.5) at the two boundary points,

$$u(0, t) = 0 \quad \text{and} \quad u(S_{\max}, t) = S_{\max} - e^{-rt}K \quad (0 \le t \le T).$$

The initial condition is given as usual by the payoff,

$$u(s, 0) = \phi(s) = \max(s - K, 0) \quad (0 < s < S_{\max}).$$

For the financial parameters, the values (1.8) are taken.

We numerically examine in this chapter the *spatial discretization error* at $t = T$ defined by

$$\varepsilon(m) = (\varepsilon_i(m))_{i=0}^{m} \quad \text{with} \quad \varepsilon_i(m) = u(s_i, T) - U_i(T) \quad (0 \le i \le m),$$

© The Author(s) 2017

K. in 't Hout, *Numerical Partial Differential Equations in Finance Explained*,
Financial Engineering Explained, DOI 10.1057/978-1-137-43569-9_5

where u denotes the exact call option value function given by the Black–Scholes formula (1.6) and U_i denotes the exact[1] semidiscrete solution function corresponding to grid point s_i. The dependence of the spatial discretization error on the number of spatial grid points is of main interest and this is therefore indicated in the notation. Let

$$e(m) = \max\{|\varepsilon_i(m)|: 0 \leq i \leq m\},$$

that is, the maximum norm of $\varepsilon(m)$. For financial applications the maximum norm is often the most relevant norm, as one wishes to uniformly control the accuracy of numerically approximated option prices.

5.1   Cell Averaging

Consider a uniform spatial grid with $m = 50$ and semidiscretization by the second-order central formulas (3.10) and (3.13) for convection and diffusion. The top of Figure 5.1 displays the spatial error components $\varepsilon_i(50)$ versus s_i. The spatial error is reasonably small throughout the spatial domain, with the largest errors occurring in a region around the strike. Consider next an increase of m by just 1 to $m = 51$. Then a similar accuracy might be expected. The bottom of Figure 5.1 shows the result, that is, $\varepsilon_i(51)$ versus s_i. Contrary to expectation, the spatial error is now substantially larger. In particular, the error increases by approximately a factor 7 around the strike. As it happens, this unfavourable outcome is not a coincidence.

Figure 5.2 displays, in double logarithmic scale, the norm of the spatial error $e(m)$ versus $1/m$ for all $10 \leq m \leq 100$. Clearly, one observes that $e(m)$ is strongly oscillating as a function of m. When m increases by 1, there is a possible increase in the spatial error with a factor of about 3.5. The reason for this undesirable behaviour lies in the nonsmoothness of the initial function at the strike. A close inspection reveals that the location of K relative to the spatial grid is correlated to the relative size of $e(m)$. In the present example, either K belongs to the spatial grid (if m is an integer multiple of three) or it has a distance of $h/3$ to the nearest spatial grid point. Figure 5.2 shows

---

[1] When computing discretization errors, the semidiscrete solution is approximated to high accuracy by applying a suitable temporal discretization method using a very small step size.

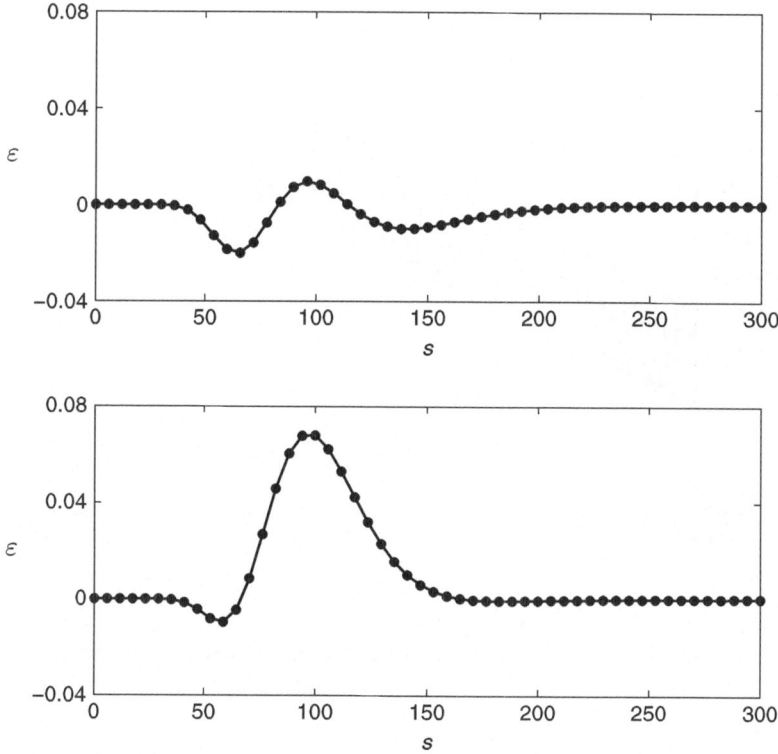

Figure 5.1 Spatial error $\varepsilon_i(m)$ versus s_i for $m = 50$ (top) and $m = 51$ (bottom). Semidiscretization on uniform grid by second-order central formulas. No cell averaging

that $e(m)$ is always relatively large in the former case, and always relatively small in the latter case. It turns out to be optimal for spatial accuracy if the strike is located exactly midway between two successive grid points. This does not occur in the present example, but could be accomplished for instance by changing S_{max}.

Here we consider application of the cell averaging technique from Chapter 4. Thus the pointwise value $\phi(s_i)$ at the grid point s_i closest to K is replaced in the initial vector $U(0)$ by the average value of ϕ over the cell $[s_{i-1/2}, s_{i+1/2}]$, see (4.10). The result is shown in Figure 5.3. One observes that the strong oscillations in $e(m)$ as a function of m have disappeared. Moreover, compared to Figure 5.2, the obtained errors are favourable. With cell averaging, $e(m)$ is seen to be approximately equal to Cm^{-2} with certain constant $C > 0$. Indeed, taking logarithms,

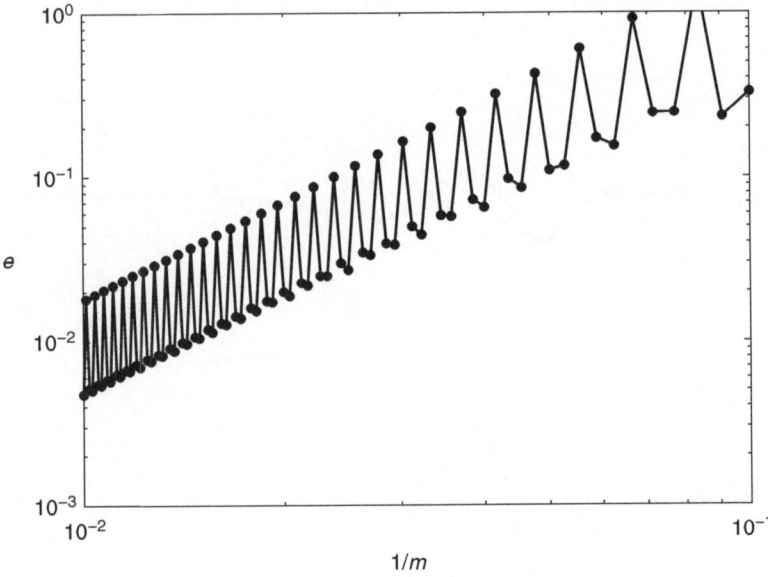

Figure 5.2 Spatial error $e(m)$ versus $1/m$ for all $10 \leq m \leq 100$. Semidiscretization on uniform grid by second-order central formulas. No cell averaging

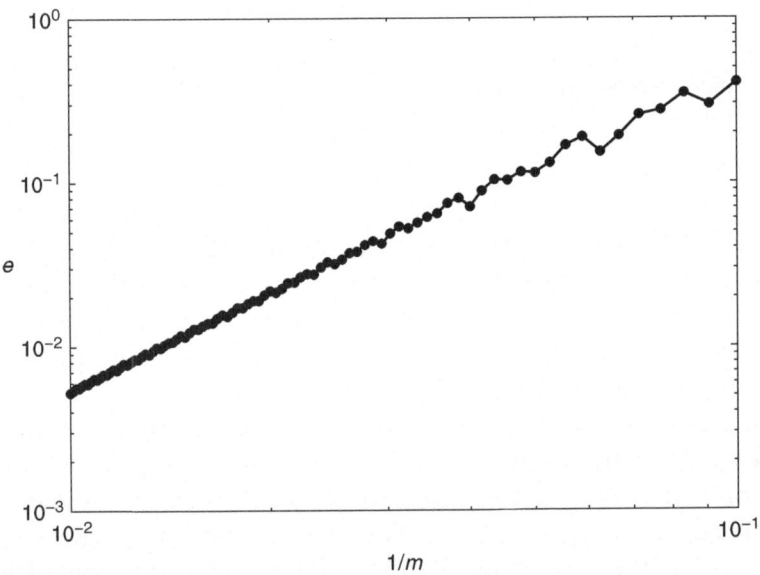

Figure 5.3 Spatial error $e(m)$ versus $1/m$ for all $10 \leq m \leq 100$. Semidiscretization on uniform grid by second-order central formulas. With cell averaging

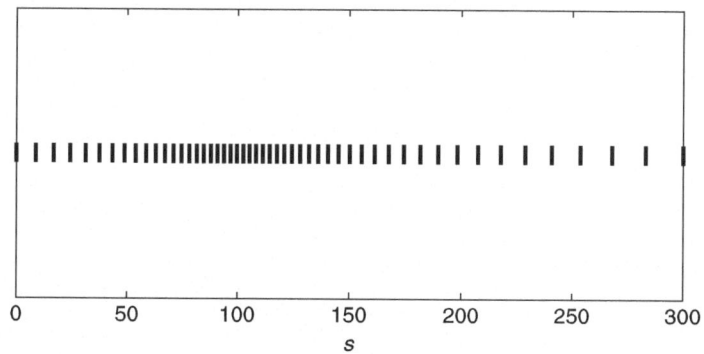

Figure 5.4 Spatial grid points corresponding to Example 4.2.1 if $m = 50$

this statement is equivalent to

$$\log e(m) \approx 2 \log (1/m) + \log C,$$

and Figure 5.3 reveals such an approximately linear relationship between $\log e(m)$ and $\log (1/m)$. Hence, a second-order convergence behaviour of the spatial discretization is attained, which is as desired.

5.2 Nonuniform Grids

As alluded to in Chapter 4, it is often natural and beneficial to use nonuniform spatial grids in financial applications. For the numerical experiments we consider the smooth, nonuniform grid constructed in Example 4.2.1. As an illustration Figure 5.4 shows the corresponding spatial grid points for $m = 50$. It is clear that the grid is densest around the strike. This forms the region of interest in practice, as asset prices do not lie far away in general from the chosen strike.

For the semidiscretization the second-order central formulas from Chapter 4 for nonuniform grids are employed and cell averaging is applied as in the previous section. Recall that for nonuniform grids there are two finite difference formulas, A and B, for the convection part, see (4.6) and (4.7). Figure 5.5 displays the spatial errors $e(m)$ versus $1/m$, where bullets correspond to formula A and squares to formula B. As a first main observation, with both formulas second-order convergence of the spatial discretization is attained. Next, the

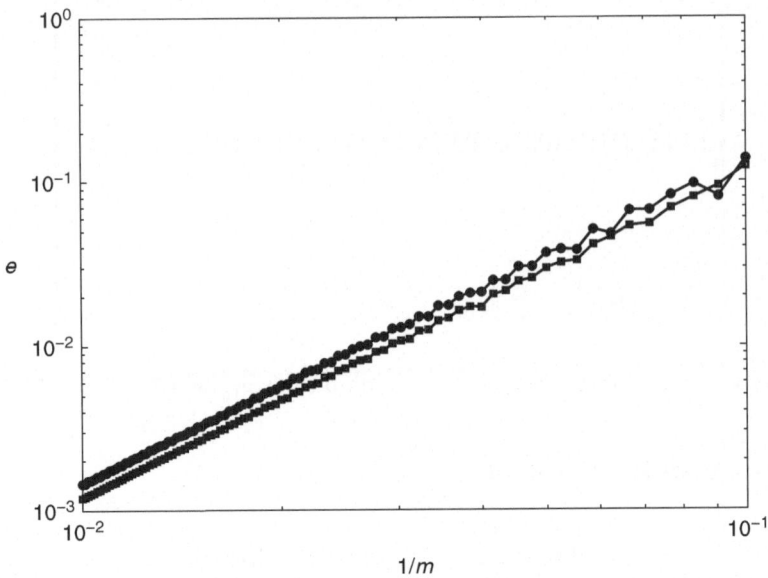

Figure 5.5   Spatial error $e(m)$ versus $1/m$ for all $10 \leq m \leq 100$. Semidiscretization on nonuniform grid by second-order central formulas. Formula A for convection: bullets. Formula B for convection: squares. With cell averaging

approximations obtained by using formula B are slightly more accurate than those obtained by using formula A. Comparing to Figure 5.3, the approximations on the nonuniform grid are with both formulas substantially more accurate than those on the uniform grid. In particular, with formula B the gain in accuracy in the present example is more than a factor of 4.

5.3   Boundary Conditions

In this section we take a closer look at the influence of the boundary condition at $s = S_{\max}$. Boundary conditions that are imposed in practice at the truncated boundaries usually introduce an error, as they are not satisfied by the exact option value function. Thus the spatial error component $\varepsilon_m(m)$ is nonzero in general.

In the situation of Sections 5.1 and 5.2, where a Dirichlet condition has been applied, a direct computation using the Black–Scholes formula (1.6) yields that $\varepsilon_m(m)$ is equal to 1.8×10^{-5} for all m. This error is always dominated in the experiments in these sections by the errors around the strike and, consequently, it does not show up in Figures 5.2, 5.3, 5.5. The same is found when the Dirichlet condition (2.5) at $s = S_{max}$ is replaced by the Neumann condition (2.6) or the linear condition (2.7) and the pertinent Figures 5.2, 5.3, 5.5 remain visually unchanged.

In the experiments so far, $10 \le m \le 100$. If the number of grid points is further increased, then $\varepsilon_m(m)$ eventually starts to dominate the spatial error over the domain $[0, S_{max}]$ and the norm of the spatial error, $e(m)$, levels off at $|\varepsilon_m(m)|$. In practice, however, only a region of interest, such as $\frac{1}{2}K < s < \frac{3}{2}K$, is of importance; compare Section 2.3. A natural alternative measure for spatial accuracy is therefore the (norm of the) *spatial discretization error on a region of interest*,

$$e^{ROI}(m) = \max\{|\varepsilon_i(m)|: 0 \le i \le m, \ \tfrac{1}{2}K < s_i < \tfrac{3}{2}K\}.$$

Figure 5.6 displays $e(m)$ and $e^{ROI}(m)$ versus $1/m$ for 50 different values m between 100 and 1000. Here the Black–Scholes PDE is semidiscretized on the nonuniform grid from Example 4.2.1, where formula B is

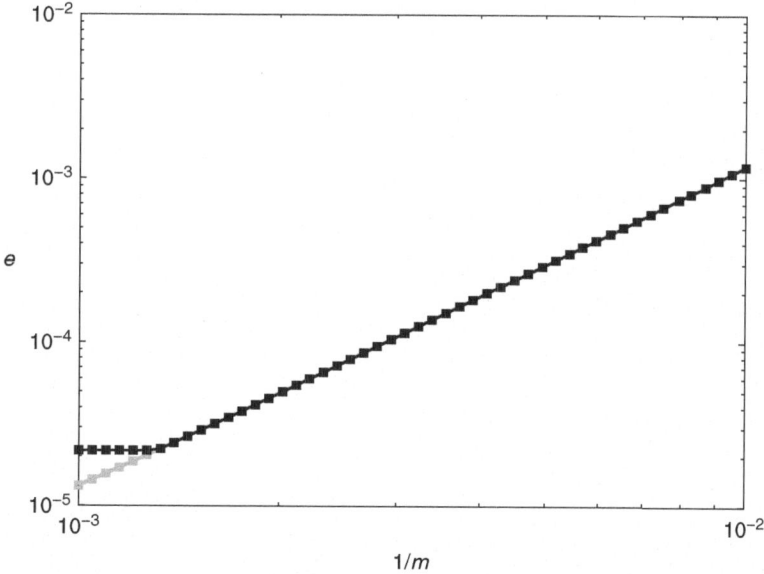

Figure 5.6 Spatial errors $e(m)$ (dark squares) and $e^{ROI}(m)$ (light squares) versus $1/m$ for $100 \le m \le 1000$. Semidiscretization on nonuniform grid by second-order central formulas. Formula B for convection. Linear boundary condition at $s = S_{max}$. With cell averaging

employed for the convection part, and cell averaging is applied. The linear boundary condition at $s = S_{\max}$ has been chosen (the results for the Dirichlet and Neumann conditions are similar). One observes that $e(m)$ (dark squares) stalls at a level of about 2×10^{-5}, which is the error at $s = S_{\max}$. On the other hand, $e^{ROI}(m)$ (light squares) continues to decrease with m according to a second-order convergence behaviour, which is as desired. The latter, favourable result is attributed to the fact that, due to the diffusion term in the Black–Scholes PDE, an error at the truncated boundary $S_{\max}$ has a negligible impact in the region of interest whenever this upper boundary is taken sufficiently large.

6

The Greeks

6.1 The Greeks

Along with the fair option values, the so-called *Greeks* are of key importance in financial practice. These quantities represent the sensitivities of an option value to changes in the underlying financial variables and parameters. A main use of Greeks is to *hedge* an option during its lifetime, that is, to reduce or eliminate risk. In mathematical terms, they are the partial derivatives of the option value with respect to its underlying variables and parameters.

In the Black–Scholes framework four well-known Greeks are

$$\text{delta}: \frac{\partial u}{\partial s}, \quad \text{gamma}: \frac{\partial^2 u}{\partial s^2}, \quad \text{vega}: \frac{\partial u}{\partial \sigma}, \quad \text{rho}: \frac{\partial u}{\partial r}.$$

These quantify sensitivities to changes in, respectively, the underlying asset price (to first and second order), the volatility and the risk-free interest rate. Notice that vega is not an actual Greek letter. For a call option one can derive from (1.6a) the following formulas (whenever $s > 0, 0 < t \le T$):

$$
\begin{array}{rll}
\text{delta:} & \mathcal{N}(d_1), & (6.1\text{a}) \\
\text{gamma:} & \mathcal{N}'(d_1)/(s\sigma\sqrt{t}), & (6.1\text{b}) \\
\text{vega:} & s\sqrt{t}\,\mathcal{N}'(d_1), & (6.1\text{c}) \\
\text{rho:} & te^{-rt}K\mathcal{N}(d_2). & (6.1\text{d})
\end{array}
$$

© The Author(s) 2017
K. in 't Hout, *Numerical Partial Differential Equations in Finance Explained*,
Financial Engineering Explained, DOI 10.1057/978-1-137-43569-9_6

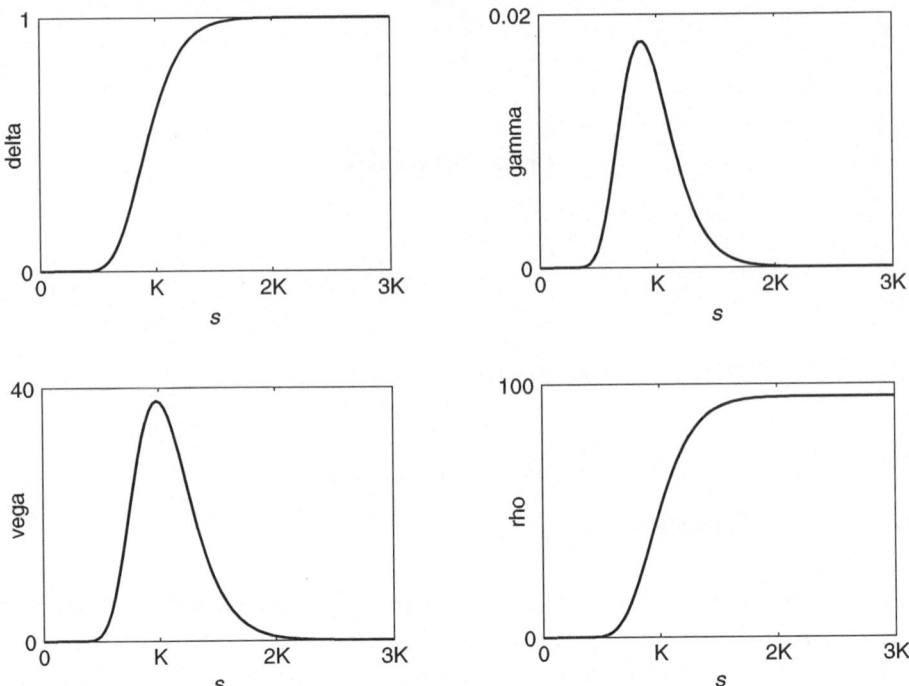

Figure 6.1   Greeks for a call option for $t = T$ and parameter set (1.8)

For a put option similar formulas are valid; these can be obtained directly from (1.6b) or by using the put-call parity (1.7). As an illustration Figure 6.1 displays the graphs of the four Greeks (6.1) in the case of parameter set (1.8) and $t = T$.

Delta and gamma appear as terms in Black–Scholes PDE, whereas vega and rho do not. Indeed, the former two Greeks concern the independent variable s, whereas the latter two concern parameters. For each Greek one can derive a convenient separate PDE. Consider for example vega and regard the option value $u = u(s, t; \sigma)$ now as a function of three variables: s, t and σ. Then vega is given by

$$v(s, t) = \frac{\partial u}{\partial \sigma}(s, t; \sigma).$$

Assuming sufficient smoothness of the function u, differentiation of all terms in the Black–Scholes PDE with respect to σ readily yields

$$v_t(s, t) = \tfrac{1}{2}\sigma^2 s^2 v_{ss}(s, t) + \sigma s^2 u_{ss}(s, t) + rsv_s(s, t) - rv(s, t) \qquad (6.2)$$

for $s > 0$ and $0 < t \leq T$. Here the dependence of u on σ has been suppressed again in the notation. The PDE (6.2) for vega is similar to the Black–Scholes PDE (1.3) for the option value. The additional term $\sigma s^2 u_{ss}(s, t)$ in (6.2) stems from application of the product rule.

For most types of options the initial and boundary conditions associated with (1.3) are independent of the volatility. Upon differentiation to σ, homogeneous initial and boundary conditions for (6.2) are then obtained.

6.2   Numerical Study

The basic approach for numerically approximating the Greeks is by straightforward use of finite differences. Here, given an option value for a set of variables and parameters, one computes one or more additional option values for perturbed (or "bumped") values of the variable or parameter of interest while keeping all others fixed, and then computes an appropriate finite difference.

If a Greek appears itself as a term in the option valuation PDE, such as delta or gamma in the Black–Scholes PDE, then a finite difference approximation is directly available at essentially no cost, since it is intrinsic to the spatial discretization of the PDE. Hence, for each option value on the spatial grid, the value of this Greek is immediately at hand.

If a Greek does not appear itself in the option valuation PDE, then in the basic approach one numerically solves this PDE for (one, two or more) additional, perturbed values of the parameter of interest and then takes an appropriate finite difference. A more natural and efficient approach in this case is to numerically solve the PDE corresponding to this Greek, simultaneously with the PDE for the option value. This requires only little extra implementation work.

For a numerical illustration consider the call option under the Black–Scholes framework with parameter values (1.8). As in Chapter 5, we truncate the spatial domain to $(0, 3K)$, select Dirichlet boundary conditions, semidiscretize on the nonuniform grid from Example 4.2.1 with second-order central formulas, using formula B for convection, and employ cell averaging. The Greeks delta and gamma are then approximated by applying the second-order central formulas (4.7) and (4.8), respectively, to the obtained option values on the spatial grid. The Greek vega is approximated by numerically solving the PDE

(6.2) along with the Black–Scholes PDE, on the same grid and in the same fashion. The Greek rho is approximated similarly to vega, by numerically solving the associated PDE. To assess the accuracy of the approximated Greeks we define, analogously to Chapter 5 for the option value and with reference to (6.1), the spatial discretization errors $\varepsilon^d(m)$, $\varepsilon^g(m)$, $\varepsilon^v(m)$, $\varepsilon^r(m)$ for delta, gamma, vega and rho, respectively, with maximum norms $e^d(m)$, $e^g(m)$, $e^v(m)$, $e^r(m)$. Figures 6.2, 6.3, 6.4, 6.5 respectively display, as squares, the latter four quantities versus $1/m$ for all $10 \leq m \leq 100$. One readily observes a nice second-order convergence behaviour in all four cases. For comparison, the results have been added to the figures where a uniform grid is employed. These are indicated by bullets. Whereas they also show a second-order convergence behaviour for all four Greeks, the accuracy is significantly lower than with the nonuniform grid, analogously to what was found in Chapter 5 for the case of the option value. Finally, we note that Figures 6.2, 6.3, 6.4, 6.5 remain visually unchanged if the other two boundary conditions, Neumann and linear, are applied for the option value function at the upper boundary.

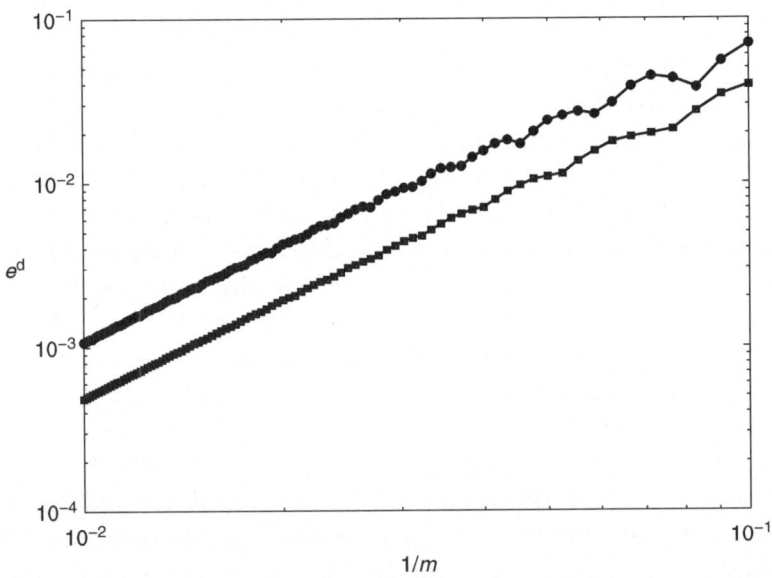

Figure 6.2 Delta spatial error $e^d(m)$ versus $1/m$ for all $10 \leq m \leq 100$. Semidiscretization by second-order central formulas. Formula B for convection. With cell averaging. Uniform grid: bullets. Nonuniform grid: squares

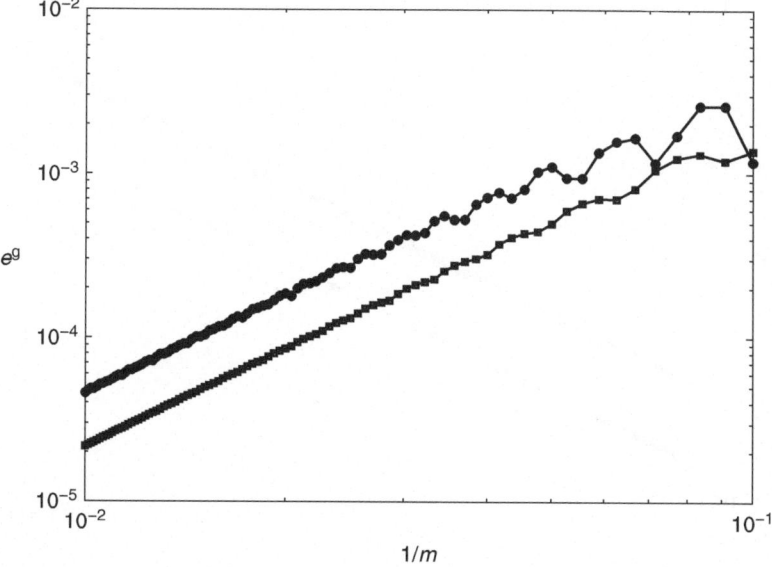

Figure 6.3   Gamma spatial error $e^g(m)$ versus $1/m$ for all $10 \leq m \leq 100$. Semidiscretization by second-order central formulas. Formula B for convection. With cell averaging. Uniform grid: bullets. Nonuniform grid: squares

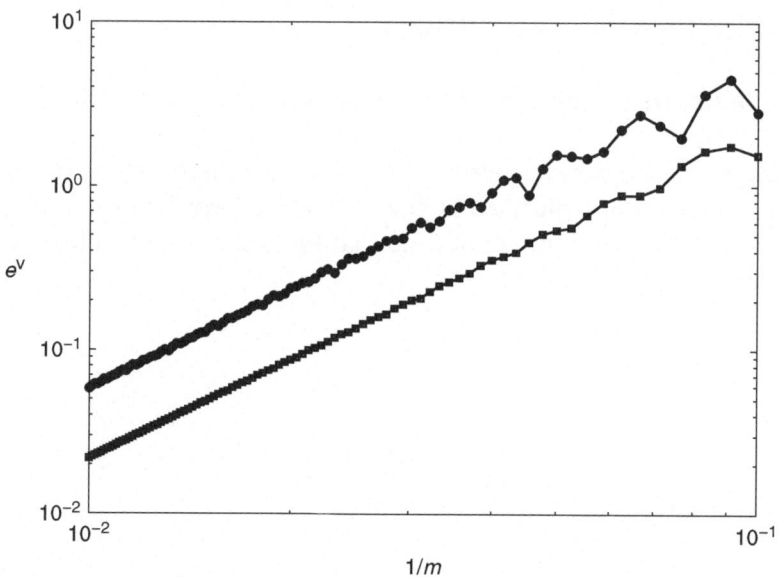

Figure 6.4   Vega spatial error $e^v(m)$ versus $1/m$ for all $10 \leq m \leq 100$. Semidiscretization by second-order central formulas. Formula B for convection. With cell averaging. Uniform grid: bullets. Nonuniform grid: squares

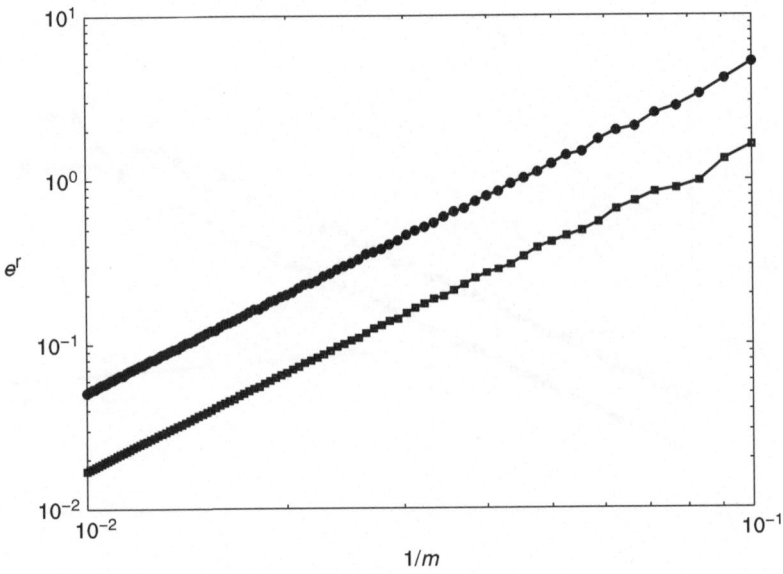

Figure 6.5   Rho spatial error $e^r(m)$ versus $1/m$ for all $10 \leq m \leq 100$. Semidiscretization by second-order central formulas. Formula B for convection. With cell averaging. Uniform grid: bullets. Nonuniform grid: squares

6.3   Notes and References

For an in-depth discussion of the Greeks in finance, see for example [47, 57, 90].

In the situation where many different Greeks need to be computed, the so-called adjoint method is very effective. Applications of the adjoint method in the context of financial PDEs are given in for example [10, 26].

7

Temporal Discretization

7.1 The θ-Methods

In this chapter we turn to the second step in the MOL approach, that is the discretization in time of the obtained semidiscrete systems; compare Chapter 3. We discuss a well-known family of temporal discretization methods in finance, the so-called θ-methods, where $\theta \in [0, 1]$ is a parameter.

Consider first an initial value problem for a general system of ODEs,

$$U'(t) = F(t, U(t)) \quad (0 < t \leq T), \quad U(0) = U_0, \tag{7.1}$$

with given function $F : [0, T] \times \mathbb{R}^\nu \to \mathbb{R}^\nu$ and given vector $U_0 \in \mathbb{R}^\nu$. Let $N \geq 1$ be any given integer and define *step size* $\Delta t = T/N$ and *temporal grid points* $t_n = n\Delta t$. Then the θ-*method* successively generates, in a one-step fashion, approximations U_n to $U(t_n)$ for $n = 1, 2, \ldots, N$ by

$$U_n = U_{n-1} + (1 - \theta)\Delta t\, F(t_{n-1}, U_{n-1}) + \theta \Delta t\, F(t_n, U_n). \tag{7.2}$$

Three notable instances of the θ-method are:

- $\theta = 0$: *forward Euler method,*
- $\theta = \frac{1}{2}$: *Crank–Nicolson method* or *trapezoidal rule,*
- $\theta = 1$: *backward Euler method.*

© The Author(s) 2017   **51**
K. in 't Hout, *Numerical Partial Differential Equations in Finance Explained,*
Financial Engineering Explained, DOI 10.1057/978-1-137-43569-9_7

If $\theta = 0$, then the method is *explicit*, and computing the approximation U_n in each time step is straightforward and cheap. If $\theta \neq 0$, then the method is *implicit* and in each time step a system of algebraic equations needs to be solved in order to determine U_n, which is obviously more expensive. As will be seen, however, implicit methods are to be preferred in general for the applications under consideration.

In this book the system of ODEs in (7.1) always stands for a semidiscrete system of the form (4.1). Then (7.2) becomes

$$U_n = U_{n-1} + (1 - \theta)\Delta t\,[AU_{n-1} + g(t_{n-1})] + \theta\Delta t\,[AU_n + g(t_n)]. \qquad (7.3)$$

The vector U_n represents the *fully discrete approximation* to the exact solution u of the pertinent PDE on the spatial grid at time level t_n. Slightly rewritten, the recurrence relation (7.3) reads

$$(I - \theta\Delta tA)U_n = (I + (1 - \theta)\Delta tA)U_{n-1} + (1 - \theta)\Delta t\,g(t_{n-1}) + \theta\Delta t\,g(t_n),$$

where I denotes the $\nu \times \nu$ identity matrix. Thus, if θ is nonzero, a linear system of equations for U_n needs to be solved, involving the matrix

$$I - \theta\Delta tA.$$

This matrix is independent of the step number n. It is therefore convenient to determine a LU factorization[1] of it once, beforehand, and employ this in all time steps to compute the vectors U_n. In many applications, the matrix A is tridiagonal and then this approach is highly efficient. Given the LU factorization of $I - \theta\Delta tA$, the number of floating-point arithmetic operations per time step to solve the linear system is directly proportional to the size ν of A, which is very favourable. This remains valid whenever A is a banded matrix with a fixed bandwidth, that is, independent of ν.

7.2 Stability and Convergence

Let $\| \cdot \|$ be any given norm on $\mathbb{R}^\nu$ and let the induced matrix norm on $\mathbb{R}^{\nu\times\nu}$ be denoted the same. Standard convergence theory for temporal

---

[1] This decomposes any given matrix into the product of a lower triangular matrix L and an upper triangular matrix U.

discretization methods applied to a given system (7.1) yields that if F is sufficiently smooth, then there exists a constant $\widehat{C}$ such that

$$\|U(t_n) - U_n\| \leq \widehat{C}(\Delta t)^q \quad \text{whenever } 1 \leq n \leq N, \ \Delta t \downarrow 0, \qquad (7.4)$$

where $q = 1$ (if $\theta \neq \frac{1}{2}$) and $q = 2$ (if $\theta = \frac{1}{2}$). Hence, the forward and backward Euler methods are first-order convergent in time and the Crank–Nicolson method is second-order convergent in time. There is a catch, however, when this temporal convergence result is used in the case of semidiscrete systems. The point is that an error constant $\widehat{C}$ is guaranteed to exist for a *fixed* ODE system, but not necessarily for a whole *class* of ODE systems. It is the latter situation that is relevant here, since one is naturally interested in letting the spatial mesh width h tend to zero, and then one arrives at an infinite class of ODE systems of unbounded size. It is not evident that a constant $\widehat{C}$ exists with the crucial property that it is *uniformly valid* for such a class of systems, under a mild condition on the step size Δt. A complete analysis of this topic is beyond the scope of this book. We shall obtain valuable insight here however by studying stability, which is of fundamental importance in its own right as well as to proving convergence.

Consider the simple scalar *test problem*

$$U'(t) = \lambda U(t) \quad (t > 0), \quad U(0) = U_0, \qquad (7.5)$$

where λ denotes any given *complex* constant. Application of the θ-method in the case of test problem (7.5) yields

$$U_n = R(\Delta t\lambda)U_{n-1} \quad (n = 1, 2, 3, \dots) \qquad (7.6)$$

with rational function $R : \mathbb{C} \to \mathbb{C}$ given by

$$R(z) = \frac{1 + (1 - \theta)z}{1 - \theta z} \quad (z \in \mathbb{C}).$$

This is called the *stability function* of the θ-method. The recurrence relation (7.6) is stable with respect to perturbations in the initial value

U_0 if and only if the modulus $|R(\Delta t \lambda)| \leq 1$. Accordingly, the *stability region* of the θ-method is defined by

$$S = \{z : z \in \mathbb{C},\ |R(z)| \leq 1\}.$$

By elementary complex analysis it is readily shown that

$$S = \begin{cases} \{z : |z + \rho| \leq \rho\} & \text{if } 0 \leq \theta < \tfrac{1}{2}, \\[2mm] \{z : \Re z \leq 0\} & \text{if } \theta = \tfrac{1}{2}, \\[2mm] \{z : |z + \rho| \geq -\rho\} & \text{if } \tfrac{1}{2} < \theta \leq 1, \end{cases}$$

where $\rho = (1-2\theta)^{-1}$. Figures 7.1, 7.2, 7.3 display the respective stability regions for the values $\theta = 0, \tfrac{1}{2}, 1$.

One easily verifies that test problem (7.5) is itself stable with respect to perturbations in U_0 if and only if $\Re \lambda \leq 0$. In view of this, a natural stability requirement on a numerical time-stepping method is that it is stable in the application to (7.5) whenever $\Re \lambda \leq 0$ and $\Delta t > 0$ (thus without any restriction on the step size). This property is called *A-stability*. It means that the left-half of the complex plane is contained

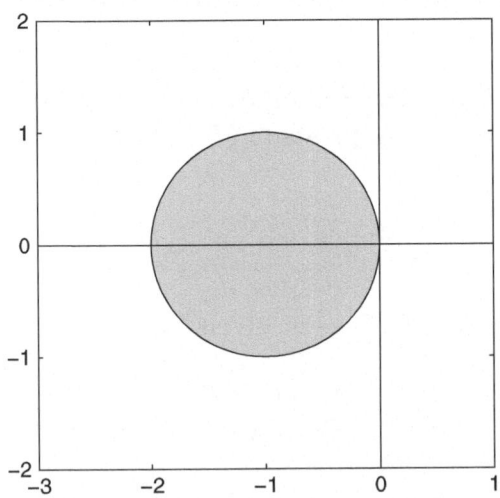

Figure 7.1   Stability region θ-method with $\theta = 0$ (shaded)

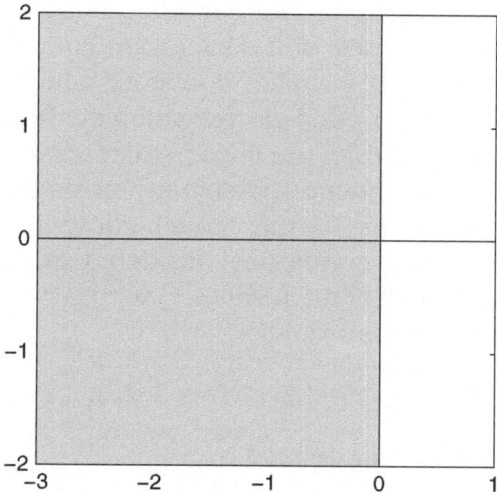

Figure 7.2   Stability region θ-method with $\theta = \frac{1}{2}$ (shaded)

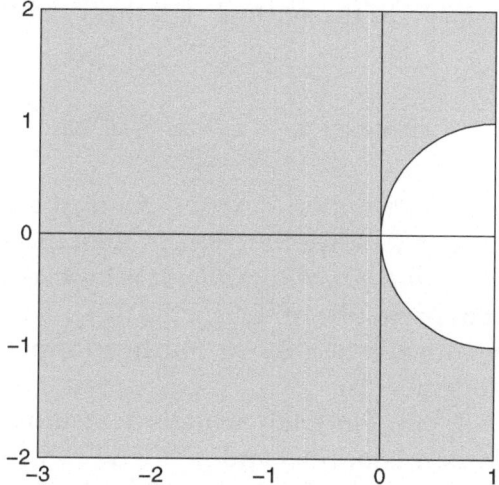

Figure 7.3   Stability region θ-method with $\theta = 1$ (shaded)

in the stability region. For the θ-method there holds

$$\{z : \Re z \leq 0\} \subset \mathcal{S} \quad \Longleftrightarrow \quad \tfrac{1}{2} \leq \theta \leq 1.$$

Hence, the Crank–Nicolson and backward Euler methods are both A-stable, whereas the forward Euler method is not. If in addition to A-stability it also holds that $R(z) \to 0$ for $z \in \mathbb{C}$ with $z \to \infty$, then the method is said to be *L-stable*. The θ-method is *L*-stable if and only if $\theta = 1$, that is, only in the case of the backward Euler method.

A-stability and *L*-stability form key concepts, which are important to the stability analysis of temporal discretization methods in applications that are far more general than the linear, scalar test problem (7.5).

We analyze in the following the stability of the θ-method applied to the semidiscretizations of the model convection and diffusion equations with periodic condition considered in Chapter 3. These semidiscretizations are all of the form (3.5) with certain $m \times m$ matrix A. Application of the θ-method yields

$$U_n = R(\Delta tA)U_{n-1} \quad (n = 1, 2, 3, \dots),$$

where $R(\Delta tA)$ is the matrix defined by

$$R(\Delta tA) = (I - \theta \Delta tA)^{-1}(I + (1 - \theta)\Delta tA),$$

provided $I - \theta \Delta tA$ is invertible. The temporal discretization is said to be *unconditionally stable* in the norm $\| \cdot \|$ if there exists a real constant M such that

$$\|R(\Delta tA)^n\| \leq M \quad \text{whenever } n \geq 1, \, n\Delta t \leq T, \, \Delta t > 0, \, h > 0. \quad (7.7)$$

If this bound on the powers of $R(\Delta tA)$ is fulfilled under a restriction $\Delta t \leq \psi(h)$ with certain positive function ψ, then the temporal discretization is said to be *conditionally stable*. It is crucial that the constant M is independent in particular of the spatial mesh width h, in line with the discussion on convergence above. Further, for practical relevance, M has to be of moderate size.

Assume $\| \cdot \| = \| \cdot \|_2$. For each semidiscretization from Chapter 3 the pertinent matrix A is normal and $A = V \Lambda V^{-1}$ with unitary matrix V and diagonal matrix $\Lambda = \text{diag}(\lambda_1, \lambda_2, \dots, \lambda_m)$. Using this, one has $R(\Delta tA)^n = VR(\Delta t\Lambda)^n V^{-1}$ and

$$\|R(\Delta tA)^n\|_2 = \|R(\Delta t\Lambda)^n\|_2 = \max_{1 \leq k \leq m} |R(\Delta t\lambda_k)|^n.$$

The following neat equivalence is then directly obtained:

$$\|R(\Delta tA)^n\|_2 \leq 1 \quad (n \geq 1) \quad (7.8)$$

$$\Updownarrow$$

$$\Delta t\lambda_k \in \mathcal{S} \quad (1 \leq k \leq m). \quad (7.9)$$

The condition (7.9) is called the *eigenvalue condition*. It only involves the step size Δt, the eigenvalues of A and the stability region S. We study this condition in the case of the different semidiscretizations from Chapter 3.

Consider first the pure model convection equation with $c < 0$ (and periodic condition) that is semidiscretized by the first-order backward formula. In Section 3.3 it was seen that all eigenvalues of the pertinent matrix A lie on a circle with midpoint $(\frac{c}{h}, 0)$ and radius $\frac{|c|}{h}$. All eigenvalues are therefore contained in the left-half of the complex plane. Hence, if $\frac{1}{2} \leq \theta \leq 1$, then (7.9) is always fulfilled whenever $\Delta t, h > 0$ and the positive result follows that the temporal discretization is unconditionally stable (in the 2-norm, with $M = 1$). If $0 \leq \theta < \frac{1}{2}$, then it is readily shown that there is conditional stability, with the step size restriction

$$\Delta t \frac{|c|}{h} \leq \rho \quad \text{or} \quad \Delta t \leq \frac{\rho}{|c|} h.$$

This is often called a *CFL condition* after Courant, Friedrichs and Lewy [15] and $c\frac{\Delta t}{h}$ is called the *Courant number*.

For the semidiscretization by the first-order forward formula, the same conclusions concerning stability as above hold, provided that $c > 0$ in this case.

Consider next the second-order central formula for convection. Then all eigenvalues of the matrix A lie on the imaginary axis. Using the equivalence of (7.8) and (7.9), one obtains again unconditional stability whenever $\frac{1}{2} \leq \theta \leq 1$.

Consider finally the semidiscretization of the pure model diffusion equation (with periodic condition) by the second-order central formula for diffusion. In this case all eigenvalues of the matrix A lie on the negative real axis and are given by

$$\lambda_k = -\frac{4d}{h^2} \sin^2(\pi k h) \quad (1 \leq k \leq m).$$

Again, the favourable result holds that the temporal discretization by the θ-method is unconditionally stable whenever $\frac{1}{2} \leq \theta \leq 1$. If $0 \leq \theta < \frac{1}{2}$, then there is conditional stability, with the step size restriction

$$\Delta t \frac{4d}{h^2} \leq 2\rho \quad \text{or} \quad \Delta t \leq \frac{\rho}{2d} h^2.$$

This upper bound on the step size is $\mathcal{O}(h^2)$, which is much more restrictive than $\mathcal{O}(h)$ as for the CFL condition. In practical applications, where the spatial mesh width can become arbitrarily small, it is too severe for an efficient discretization.

Concerning the full model equation (2.3) with periodic condition it holds that the θ-method is unconditionally stable whenever $\frac{1}{2} \leq \theta \leq 1$ and the convection and diffusion parts are semidiscretized by their second-order central finite difference formulas. In this case the temporal convergence bound (7.4) holds in the scaled Euclidean norm (3.16) with order q as before and error constant $\widehat{C}$ independent of the spatial mesh width h, as desired, if u_0 is sufficiently smooth; compare [49]. These results remain valid if the convection part is semidiscretized by the first-order backward (forward) formula and $c < 0$ ($c > 0$).

7.3   Maximum Norm and Positivity

Obtaining for full discretizations favourable stability results in the maximum norm $\|\cdot\|_\infty$ is often more difficult than in the (scaled) Euclidean norm. For the backward Euler method, a useful bound on $\|R(\Delta tA)\|_\infty = \|(I - \Delta tA)^{-1}\|_\infty$ can be derived if the condition (4.11) on the matrix A of the form (4.2) holds with $r \geq 0$, which is valid for the mixed central/upwind discretization from Section 4.4. We first note that the inverse of $I - \Delta tA$ exists. Consider then any given vector x of size ν and

$$y = (I - \Delta tA)^{-1}x.$$

Let index $i \in \{1, 2, \ldots, \nu\}$ be such that $\|y\|_\infty = |y_i|$. Since

$$x_i = (1 - \Delta t\alpha_i)y_i - \Delta t\beta_i y_{i-1} - \Delta t\gamma_i y_{i+1},$$

with $\beta_1 = \gamma_\nu = 0$ and $y_0 = y_{\nu+1} = 0$, it follows using (4.11) that

$$
\begin{aligned}
|x_i| &\geq |1 - \Delta t\alpha_i||y_i| - \Delta t|\beta_i||y_{i-1}| - \Delta t|\gamma_i||y_{i+1}| \\
&\geq (1 - \Delta t\alpha_i)|y_i| - \Delta t\beta_i|y_i| - \Delta t\gamma_i|y_i| \\
&= (1 - \Delta t\alpha_i - \Delta t\beta_i - \Delta t\gamma_i)|y_i| \\
&\geq (1 + r\Delta t)|y_i|.
\end{aligned}
$$

Hence,

$$\|y\|_\infty = |y_i| \le \frac{1}{1+r\Delta t}\,|x_i| \le \frac{1}{1+r\Delta t}\,\|x\|_\infty.$$

As the vector x is arbitrary, one obtains the stability bound

$$\|(I - \Delta t A)^{-1}\|_\infty \le \frac{1}{1+r\Delta t} \quad \text{for all } \Delta t > 0. \tag{7.10}$$

Since this upper bound is at most 1, the backward Euler method is unconditionally contractive in the maximum norm, which is a very strong property.

We subsequently show that it is also positivity preserving. To this purpose, consider any given vector $x > 0$ (componentwise) and regard y as a function of $\Delta t > 0$, thus, $y = y(\Delta t)$. First, the vector function y is continuous. Next, it is strictly positive for all sufficiently small Δt. Suppose there exist a step size $\Delta t^* > 0$ and an index $i \in \{1, 2, \dots, \nu\}$ such that $y^* = y(\Delta t^*)$ satisfies

$$y_i^* = 0 \quad \text{and} \quad y_j^* \ge 0 \quad \text{(whenever } j \ne i).$$

Then, using (4.11),

$$0 < x_i = (1 - \Delta t \alpha_i) y_i^* - \Delta t \beta_i y_{i-1}^* - \Delta t \gamma_i y_{i+1}^* \le 0,$$

which is a contradiction. This implies that $y(\Delta t) > 0$ for all $\Delta t > 0$. Since $x > 0$ is arbitrary, it follows that

$$(I - \Delta t A)^{-1} \ge 0 \quad \text{for all } \Delta t > 0. \tag{7.11}$$

Consequently, the backward Euler method is unconditionally positivity preserving: for any $N \ge 1$, the approximations U_n defined by (7.3) with $\theta = 1$ satisfy

$$U_0 \ge 0 \quad \text{and} \quad g(t) \ge 0 \ (0 \le t \le T) \quad \Longrightarrow \quad U_n \ge 0 \ (1 \le n \le N).$$

Under the condition (4.11) the favourable result can also be proved that the backward Euler method is free from spurious oscillations whenever $\Delta t > 0$.

The above strong properties of the backward Euler method and the ideas of proof extend far beyond the present application.

For the θ-methods with $\theta < 1$ the same properties as above can be shown, provided a strict condition on the step size holds.

7.4 Notes and References

The family of θ-methods belongs to two general classes of temporal discretization methods, the Runge–Kutta methods and the linear multistep methods. For these two classes of methods an ample theory is available that is relevant to their application to semidiscrete systems, see for example the books [9, 31, 32]. Here initial value problems for ODE systems obtained from semidiscretization of diffusion-dominated PDEs (2.1) are so-called *stiff* problems. For the effective numerical solution of stiff problems well-selected implicit methods are preferable over explicit methods. Here A-stability forms a pivotal property. Besides the θ-methods with $\frac{1}{2} \leq \theta \leq 1$ many other A-stable methods are known; compare the references above.

In the book [49] a comprehensive convergence analysis is given of Runge–Kutta methods and linear multistep methods when applied to ODE systems stemming from semidiscretization by finite difference formulas of time-dependent convection-diffusion-reaction equations. The analysis follows the symbolic law

$$\textit{stability and consistency} \implies \textit{convergence}.$$

Assuming consistency and some additional conditions, the reverse implication is also true. This is the content of the famous Lax–Richtmyer theorem, see for example [83] for details.

The stability analysis in Section 7.2 relies upon the matrix A being normal. In practical applications, however, it is usually nonnormal. The equivalence of (7.8) and the eigenvalue condition (7.9) is then not valid anymore. The latter condition is necessary, but not sufficient in general, for the former. Still it is often applied also in this situation to claim stability, but due care needs to be taken. Many rigorous stability results pertinent to nonnormal matrices A are given in for example the texts [82, 87].

Stability bounds for general norms, in particular the important maximum norm, are included in the references above, notably [82]. Ample results on monotonicity properties for temporal discretization methods, such as those considered in Section 7.3, are given in for example [49]. The prevention of spurious oscillations in the numerical solution of financial option valuation PDEs was first addressed in [96].

8

Numerical Study: Time

In the following we examine through numerical experiments the stability and convergence of the temporal discretization by the θ-methods in the valuation of a call option under the Black–Scholes framework. Continuing the numerical example from Chapters 5 and 6, the financial parameter set (1.8) is taken, the spatial domain is truncated to $(0, 3K)$ and Dirichlet conditions (1.5) and (2.5) at the boundaries are chosen. Next, spatial discretization is performed on the nonuniform grid specified in Example 4.2.1 with second-order central formulas for convection and diffusion, applying formula B for convection. Cell averaging is always employed in this chapter, so that the pointwise payoff value in the initial vector $U(0)$ at the grid point s_i closest to the strike K is replaced by its cell averaged value. This all together leads to an initial value problem for a semidiscrete system of the form (4.1) with size $\nu = m - 1$.

8.1 Explicit Method

First consider application of the forward Euler method. Set $m = 50$ for the spatial grid and choose $N = 75$ and $N = 80$ time steps. Figure 8.1 displays the two obtained fully discrete approximations at $t = T$. The top graph shows the result for $N = 75$. Clearly, this numerical solution is very poor, with excessively large positive and large negative values. The bottom graph shows the result for $N = 80$. This forms a visually

© The Author(s) 2017
K. in 't Hout, *Numerical Partial Differential Equations in Finance Explained*,
Financial Engineering Explained, DOI 10.1057/978-1-137-43569-9_8

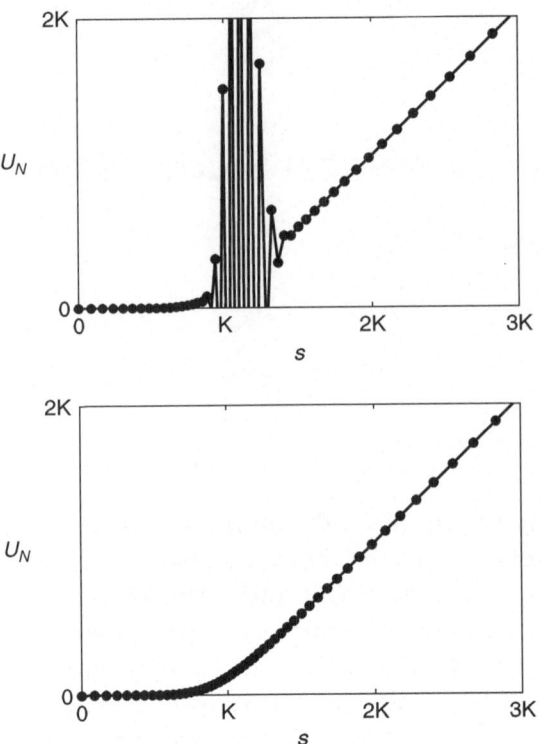

Figure 8.1   Fully discrete approximation of call option value function for $t = T$ obtained with the forward Euler method if $N = 75$ (top) and $N = 80$ (bottom)

fine numerical solution; compare Figure 1.2 for the graph of the exact solution. Selecting smaller numbers of time steps than $N = 75$ leads to similar incorrect results as in the top of Figure 8.1, whereas taking larger numbers than $N = 80$ yields decent results. Hence, there appears to be a critical, minimal number of steps that is required, equal to about $N = 80$. Considering next $m = 100, 200, 400$ numerical experiments reveal critical numbers of time steps approximately equal to $N = 330, 1330, 5350$, respectively. These numbers grow fast and appear to be directly proportional to m^2.

To gain insight into the above observation we take a look at the eigenvalue condition (7.9), with m replaced by ν. Computation in Matlab of the eigenvalues $\lambda_1, \lambda_2, \ldots, \lambda_\nu$ of the pertinent $\nu \times \nu$ matrix A indicates that, for each value m considered, they all lie on the negative real axis and, with the corresponding value N found above and $\Delta t = T/N$, it holds that

$$\min_{1\leq k\leq v} \Delta t\lambda_k \in [-2.02,-1.99].$$

The intersection of the stability region $\mathcal{S}$ of the forward Euler method with the negative real axis is the interval $[-2, 0]$. Hence, for each m, the corresponding value N above is very close to the minimal value N for which the eigenvalue condition (7.9) is fulfilled. If the number of time steps is smaller than this critical value, then an unstable behaviour of the forward Euler method is to be expected, which is indeed what the numerical experiments indicate. In particular, if $m = 50$ and $N = 75$, then $\min_k \Delta t\lambda_k = -2.15$. This point clearly lies outside of $\mathcal{S}$, explaining the poor fully discrete approximation in this case.

Subsequent numerical experiments suggest that there is a stable behaviour of the forward Euler method whenever the number of time steps is larger than the critical value. Even though the present matrix A is nonnormal, fulfilment of (7.9) thus appears to guarantee conditional stability. As alluded to before, the relevant step size restriction is of the form $\Delta t = \mathcal{O}(m^{-2})$. This is of the same kind as obtained in Chapter 7 in the case of the pure model diffusion equation with periodic condition. For practical applications, where efficiency is crucial, this step size restriction is too severe.

8.2 Implicit Methods

In view of the unfavourable experience with the forward Euler method, we consider in this section numerical experiments with two implicit methods, backward Euler and Crank–Nicolson. It turns out that with these two methods always a visually fine numerical solution is obtained, for any given number of time steps. This forms a striking, positive difference with the case of the forward Euler method. Relatedly, since the backward Euler and Crank–Nicolson methods are both A-stable, condition (7.9) is now always fulfilled whenever the eigenvalues of the matrix A lie in the left-half of the complex plane.

To study the convergence behaviour of the two implicit methods, we consider the *temporal discretization error* at $t = T = N\Delta t$ defined by

$$\widehat{\varepsilon}(\Delta t;m) = (\widehat{\varepsilon}_i(\Delta t;m))_{i=0}^{m} \quad \text{with} \quad \widehat{\varepsilon}_i(\Delta t;m) = U_i(T) - U_{N,i} \quad (0 \leq i \leq m),$$

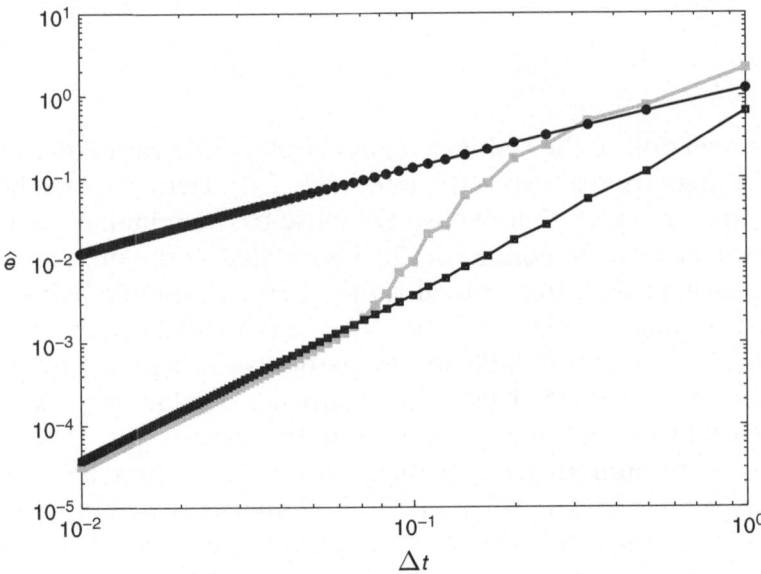

Figure 8.2   Temporal error $\widehat{e}(\Delta t;50)$ versus Δt for all $1 \leq N \leq 100$. Backward Euler (dark bullets), Crank–Nicolson (light squares), Crank–Nicolson with damping (dark squares)

and maximum norm

$$\widehat{e}(\Delta t;m) = \max\{|\widehat{\varepsilon}_i(\Delta t;m)|: 0 \leq i \leq m\}.$$

Figures 8.2 and 8.3 display for $m = 50$ and $m = 200$, respectively, the values $\widehat{e}(\Delta t;m)$ versus Δt in double logarithmic scale for all $1 \leq N \leq 100$. The results for backward Euler are shown by dark bullets and those for Crank–Nicolson by light squares. The dark squares are discussed below.

For the backward Euler method one observes that the (norm of the) temporal discretization errors are all bounded from above by a moderate constant and decrease monotonically as N increases. The slope of the line through the results for backward Euler is 1.0, which indicates that $\widehat{e}(\Delta t;m)$ is approximately equal to $\widehat{C}\Delta t$ with constant $\widehat{C}$ independent of N. This corresponds to a first-order convergence behaviour, as expected. Comparing next the lines for $m = 50$ and $m = 200$ in Figures 8.2 and 8.3 for backward Euler, no vertical shift is observable, indicating that $\widehat{C}$ *is independent of* m, as desired; see Section 7.2. Additional numerical experiments with larger values m support this main conclusion.

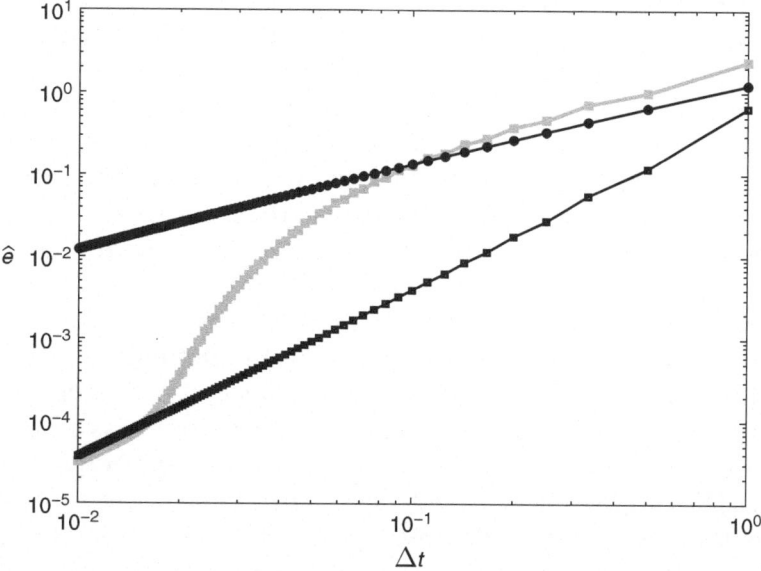

Figure 8.3   Temporal error $\hat{e}(\Delta t;200)$ versus Δt for all $1 \le N \le 100$. Backward Euler (dark bullets), Crank–Nicolson (light squares), Crank–Nicolson with damping (dark squares)

For the Crank–Nicolson method the temporal discretization errors are also all bounded from above by a moderate constant and decrease monotonically as N increases. Furthermore, they are smaller in general than those obtained for the backward Euler method. The observed convergence behaviour for Crank–Nicolson is odd, however. Three separate regions in the step size domain can be distinguished. For large Δt there seems to be approximately first-order convergence, then there is a transition region where the temporal errors drop fast, and finally for small Δt second-order convergence is achieved. Next, comparing Figures 8.2 and 8.3, the region of step sizes where approximately first-order convergence holds for Crank–Nicolson expands significantly when the number of spatial grid points is increased, which is confirmed by additional experiments with larger values m. In practical applications, these are undesirable features.

The peculiar convergence behaviour of the Crank–Nicolson method can be explained from the facts that the initial function is nonsmooth and that, even though the method is A-stable, its stability function R has a modulus equal to 1 at infinity. A brief sketch of this is as follows. First, due to the nonsmoothness of the initial function, and in spite of cell averaging being applied, relatively large temporal discretization errors occur in the initial time steps, where the components $\widehat{\varepsilon}_i(\Delta t;m)$ spike near the strike, that is, when $s_i \approx K$. Next, because $|R(\infty)| = 1$, the (sizeable) contributions to these errors of the eigenvectors corresponding to the dominant eigenvalues λ_k of A (here the eigenvalues that are large negative) are not adequately damped in the subsequent time steps, unless the step size is taken sufficiently small. We mention that these contributions are referred to as *high-frequency modes* in a Fourier terminology.

There exists a simple and effective remedy for the undesirable convergence behaviour of the Crank–Nicolson method. It consists of replacing the very first time step by two substeps with step size $\Delta t/2$ using the backward Euler method. The latter method is L-stable and therefore provides immediate damping of the high-frequency error modes. This approach is called *backward Euler damping* or *Rannacher time stepping*. In Figures 8.2 and 8.3 the results obtained by Crank–Nicolson with damping are indicated by dark squares. The improvement is clear and as desired. First, the temporal discretization errors are always either approximately equal to or substantially smaller than those obtained by Crank–Nicolson without damping. Next, their norms $\widehat{e}(\Delta t;m)$ are now approximately equal to $\widehat{C}(\Delta t)^2$ with $\widehat{C}$ independent of N and m. Hence, second-order convergence in time is recovered, uniformly in m.

In applications where the initial function is highly nonsmooth, such as Dirac delta initial data, it is recommended to replace the first two time steps by four half-steps of the backward Euler method.

We subsequently study the *total discretization error*, which is of primary importance in practice. It is equal to the sum of the spatial and temporal discretization errors, that is $\varepsilon(m) + \widehat{\varepsilon}(\Delta t;m)$, and has maximum norm

$$E(\Delta t;m) = \max\{|u(s_i, T) - U_{N,i}|: 0 \leq i \leq m\}.$$

In Chapter 5 second-order convergence of the current spatial discretization was obtained. It is thus natural to select for the temporal discretization the Crank–Nicolson method with damping, for which second-order convergence in time, uniformly in m, has been found. Then *the number of time steps N can be chosen directly proportional to m* so that the spatial and temporal discretization errors decrease at the same rate. As an illustration, Figure 8.4 displays $E(\Delta t; m)$ versus $1/m$ with $N = \lceil m/5 \rceil$ and $10 \leq m \leq 1000$ for the backward Euler method (dark bullets), the Crank–Nicolson method without damping (light squares) and the Crank–Nicolson method with damping (dark squares). One readily observes that the total errors in the case of Crank–Nicolson with damping are $\mathcal{O}(m^{-2})$. Without damping, they behave erratically, and a substantial loss of accuracy occurs for larger m. The total errors in the case of backward Euler are $\mathcal{O}(m^{-1})$. These observations are all consistent with the foregoing results. We conclude that of the methods under consideration the Crank–Nicolson method with damping is preferable in terms of accuracy and efficiency.

Numerical experiments for each of the Greeks delta, gamma, vega, rho also reveal an $\mathcal{O}(m^{-2})$ behaviour of the total errors in the case of Crank–Nicolson with damping, provided that for gamma four damping substeps are used and the total error is considered on the region of interest, see Section 5.3. Further, Figures 8.2–8.4 remain virtually unchanged if formula A is chosen for convection or if the Neumann or linear boundary condition is taken at the upper boundary.

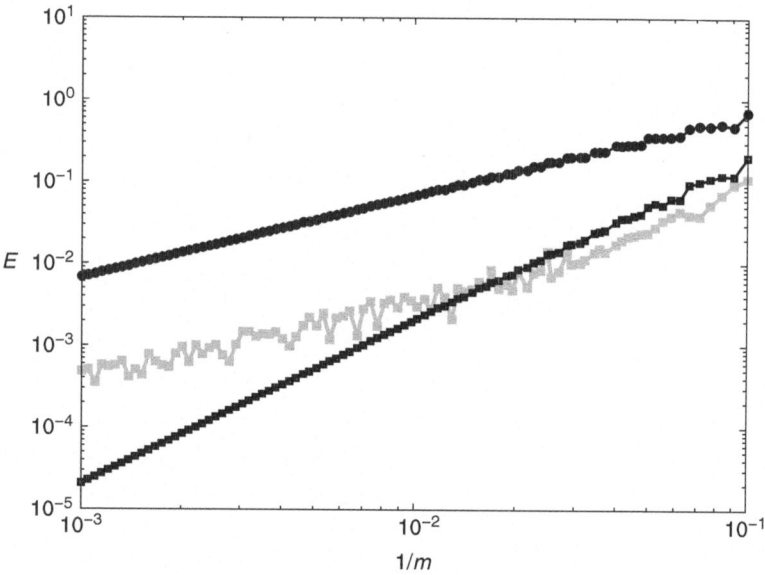

Figure 8.4   Total error $E(\Delta t; m)$ versus $1/m$ with $N = \lceil m/5 \rceil$ for $10 \leq m \leq 1000$. Backward Euler (dark bullets), Crank–Nicolson (light squares), Crank–Nicolson with damping (dark squares)

8.3 Notes and References

Rigorous, theoretical stability bounds for the spatial and temporal discretizations of the Black–Scholes PDE under consideration have been derived in [39].

Rannacher time stepping is due to [72]. It was first analyzed for applications in computational finance by [25, 71] and, along with subsequent work by these authors, other initial references to this damping procedure are [27, 89].

9

Cash-or-Nothing Options

As a first step towards the numerical PDE valuation of more advanced types of financial options, we consider the cash-or-nothing call option, see Chapter 4. This option is relatively simple, but its payoff function has the property that it is discontinuous at the strike, see (4.9).

Options having discontinuous payoffs are referred to as *digital options* or *binary options*. Their numerical valuation, and notably the approximation of their Greeks delta and gamma, requires careful attention.

The fair value $u(s, t)$ of a cash-or-nothing call option under the Black–Scholes framework is given by

$$u(s,t) = e^{-rt}D\mathcal{N}(d_2) \quad \text{(for } s > 0,\ 0 < t \leq T\text{)}, \tag{9.1}$$

with d_2 given in the Black–Scholes formula (1.6). As a numerical example we take

$$K = 100,\ D = 100,\ T = 0.5,\ r = 0.03,\ \sigma = 0.40. \tag{9.2}$$

Figure 9.1 displays the corresponding graph of u in the (s, t)-domain $[0, 3K] \times [0, T]$. The discontinuity in the payoff function is clearly observable at $t = 0$.

For the numerical PDE valuation we consider the same spatial discretization as in Chapter 8 for a call option, except now with the

© The Author(s) 2017 **69**
K. in 't Hout, *Numerical Partial Differential Equations in Finance Explained*,
Financial Engineering Explained, DOI 10.1057/978-1-137-43569-9_9

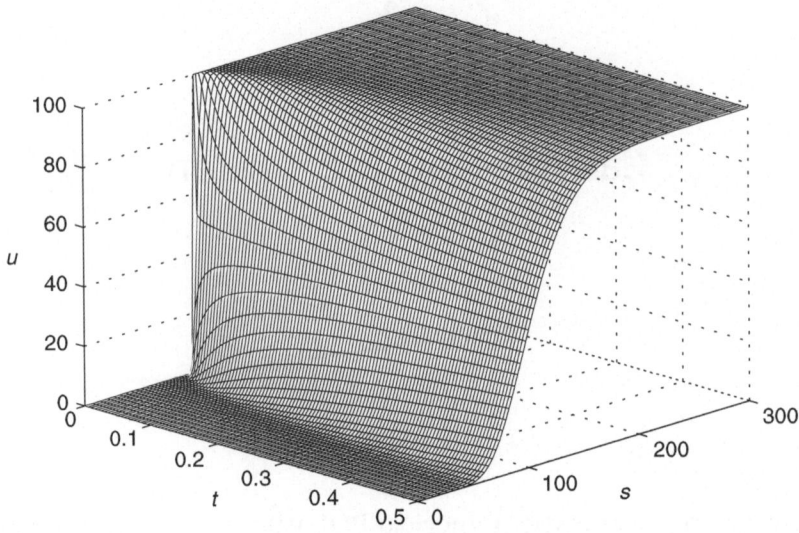

Figure 9.1   Exact cash-or-nothing call option value function on $[0, 3K] \times [0, T]$ with parameter set (9.2)

payoff function (4.9) for the initial condition and a different Dirichlet condition at the upper boundary,

$$u(S_{\max}, t) = e^{-rt} D \quad (0 \leq t \leq T).$$

For the temporal discretization of the obtained semidiscrete system (4.1), the Crank–Nicolson method is applied.

Analogously to Chapter 8 we study the norm of the total discretization error $E^{ROI}(\Delta t; m)$ versus $1/m$ with $\Delta t = T/N$, $N = \lceil m/5 \rceil$ and $10 \leq m \leq 1000$, where the error is now restricted to the region of interest given by $\frac{1}{2}K < s < \frac{3}{2}K$. We are interested in the effects that the techniques of cell averaging (Section 4.3) and backward Euler damping (Section 8.2) have in the numerical valuation. Figure 9.2 displays the total errors in three cases: with cell averaging but no damping (dark bullets), with cell averaging and damping (dark squares), and no cell averaging but with damping (light squares). Here damping is performed by two substeps of backward Euler. From Figure 9.2 it is clear that the two techniques are both necessary to arrive at accurate approximations and a regular convergence behaviour. If the two techniques are simultaneously employed, then second-order convergence

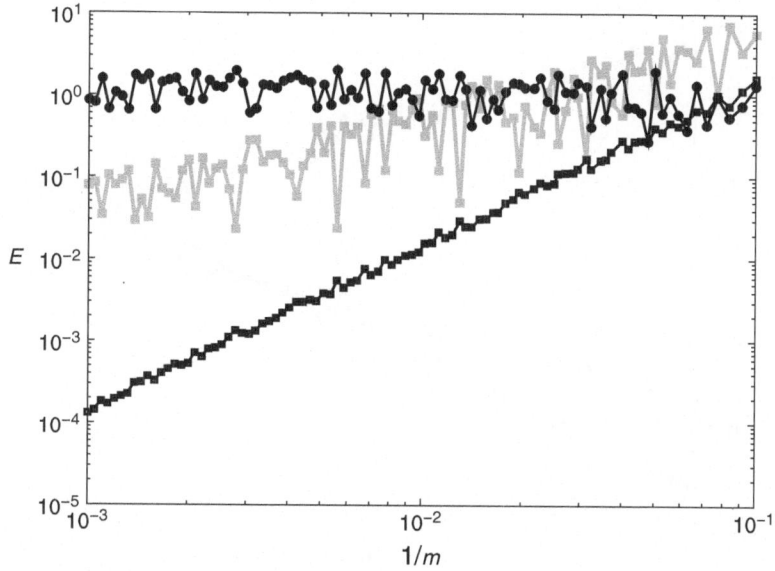

Figure 9.2 Cash-or-nothing call option with parameter set (9.2). Total error $E^{ROI}(\Delta t; m)$ versus $1/m$ with $N = \lceil m/5 \rceil$ for $10 \leq m \leq 1000$. Crank–Nicolson method. Cell averaging without damping (dark bullets). Cell averaging with damping using two substeps (dark squares). No cell averaging but with damping using two substeps (light squares)

is achieved, that is, the total errors are $\mathcal{O}(m^{-2})$. If one of the two techniques is omitted, then the errors are large in general and, moreover, behave erratically as a function of m.

For the Greeks delta and gamma the accurate numerical approximation is more challenging than that for the option value itself, due to an increased lack of smoothness near $t = 0$. The following exact formulas easily follow from (9.1):

$$\text{delta:} \quad e^{-rt} D \mathcal{N}'(d_2)/(s\sigma\sqrt{t}),$$
$$\text{gamma:} \quad -e^{-rt} D d_1 \mathcal{N}'(d_2)/(s^2\sigma^2 t),$$

for $s > 0$ and $0 < t \leq T$. These formulas readily reveal that delta and gamma are both unbounded at the strike if $t \downarrow 0$, as one could expect. Analogously as above we examine the norm of the pertinent total discretization errors $E^{d,ROI}(\Delta t; m)$ and $E^{g,ROI}(\Delta t; m)$ versus $1/m$ where $N = \lceil m/5 \rceil$. Four cases are considered: cell averaging but no damping (dark bullets), cell averaging and damping using two substeps (dark squares), cell averaging and damping using four substeps

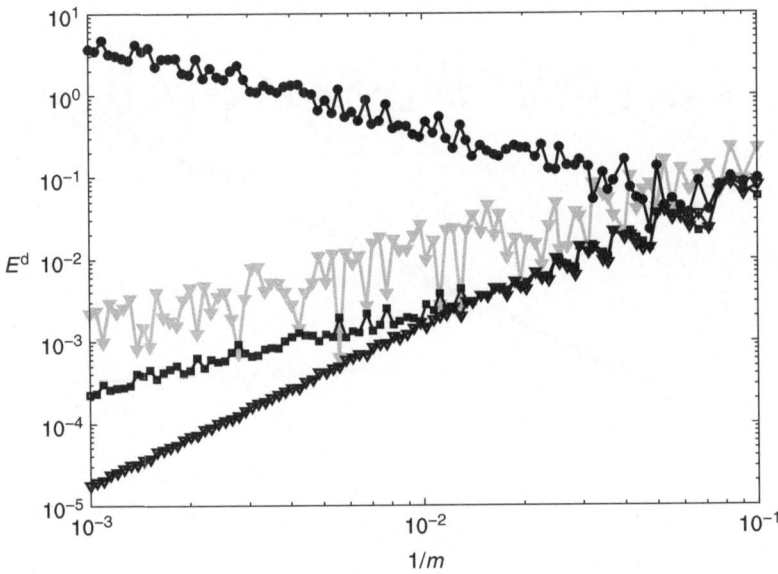

Figure 9.3   Cash-or-nothing call option delta with parameter set (9.2). Total error $E^{d,ROI}(\Delta t;m)$ versus $1/m$ with $N = \lceil m/5 \rceil$ for $10 \leq m \leq 1000$. Crank–Nicolson method. Cell averaging and: no damping (dark bullets), damping using two substeps (dark squares) and damping using four substeps (dark triangles). No cell averaging but with damping using four substeps (light triangles)

(dark triangles), and no cell averaging but with damping using four substeps (light triangles). Figures 9.3 and 9.4 display the total errors for, respectively, delta and gamma. It is clear that only when cell averaging and damping with four substeps are both applied, second-order convergence for delta and gamma is attained. If just two substeps of backward Euler are used for damping, then only first-order convergence is found for delta, whereas for gamma convergence is absent in this case. Without damping, the errors for delta and gamma both diverge as m increases. Finally, without cell averaging, the errors are mostly much larger than with cell averaging and the error behaviour is erratic.

A *cash-or-nothing put option* has the payoff

$$\phi(s) = \begin{cases} D & \text{for } s < K, \\ 0 & \text{for } s > K. \end{cases}$$

In a general setting, which includes the Black–Scholes framework, the sum of the fair values of corresponding cash-or-nothing call and put

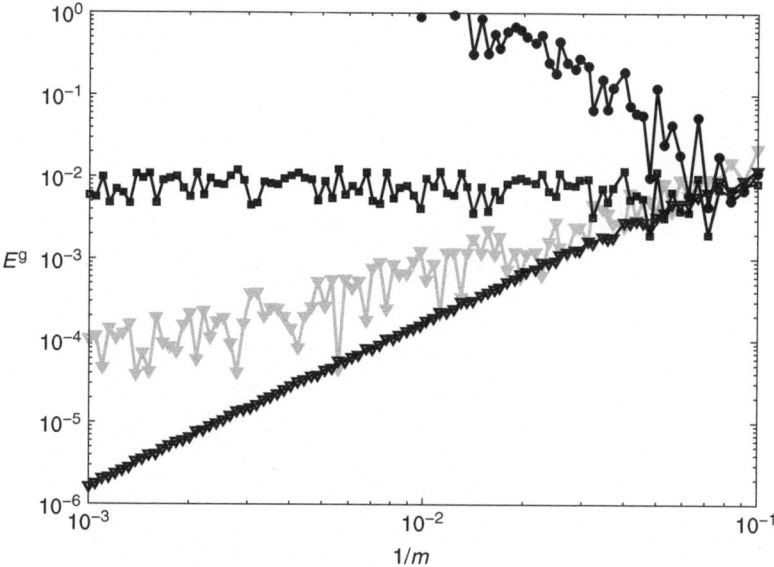

Figure 9.4 Cash-or-nothing call option gamma with parameter set (9.2). Total error $E^{g,ROI}(\Delta t;m)$ versus $1/m$ with $N = \lceil m/5 \rceil$ for $10 \leq m \leq 1000$. Crank–Nicolson method. Cell averaging and: no damping (dark bullets), damping using two substeps (dark squares) and damping using four substeps (dark triangles). No cell averaging but with damping using four substeps (light triangles)

options is equal to $e^{-rt}D$, a parity relation. Relatedly, the same conclusions concerning the total errors in the case of a cash-or-nothing put option are obtained as above.

10

Barrier Options

Barrier options are among the most common exotic options used in practice. They are *path-dependent*, that is, their payoffs depend not only on the underlying asset prices at maturity, but also on those prior to maturity. We consider here the example of a *down-and-out put option*. This option gives the holder the right to sell the underlying asset for strike price K at maturity time T provided that the asset price does not fall below a prescribed barrier H during the lifetime $[0, T]$ of the option; otherwise the option is worthless. If the barrier is breached, then one speaks of a *knock-out* event.

A main reason for the popularity of barrier options is that they are often much cheaper than their direct counterparts without barriers. A semi-closed analytical formula for the fair value $u(s, t)$ of a down-and-out put option under the Black–Scholes framework is known; it is provided in Appendix C. If $t = 0$, then the fair value equals $\max(K-s, 0)$ for $s > H$ and it is zero for $s \leq H$.

Figure 10.1 displays the graph of u on $[H, 3K] \times [0, T]$ for the sample parameter set

$$K = 100, \ H = 75, \ T = 1, \ r = 0.06, \ \sigma = 0.30. \qquad (10.1)$$

In the following we study the numerical PDE valuation of the down-and-out put option. The spatial discretization is performed analogously to that in Chapters 8 and 9. The only modifications in the semidiscretization with respect to that for a standard put option are that the spatial

K. in 't Hout, *Numerical Partial Differential Equations in Finance Explained*,
Financial Engineering Explained, DOI 10.1057/978-1-137-43569-9_10

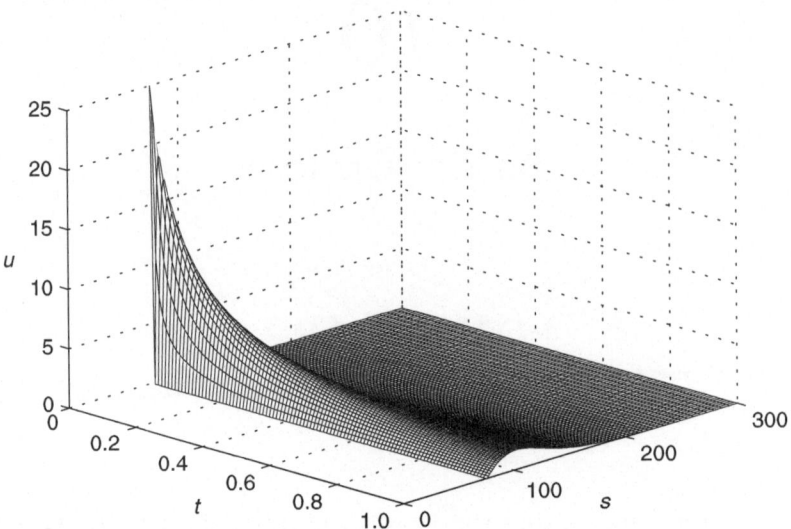

Figure 10.1 Exact down-and-out put option value function on $[H, 3K] \times [0, T]$ with parameter set (10.1)

domain is now given by

$$(S_{\min}, S_{\max}) = (H, 3K)$$

and that a homogeneous Dirichlet condition at the lower boundary $s = H$ holds since the fair value of a down-and-out put is always zero there. The fair value at $t = 0$ (above) determines as usual the initial condition. For the temporal discretization of the resulting semidiscrete system (4.1), the Crank–Nicolson method is applied.

We consider the norm of the total discretization error $E^{ROI}(\Delta t; m)$ versus $1/m$ with $\Delta t = T/N$, $N = \lceil m/5 \rceil$ and $10 \leq m \leq 1000$. The region of interest is now taken as $H < s < \frac{3}{2}K$. In Figure 10.2 the total errors are displayed for three cases: with cell averaging but no damping (dark bullets), with cell averaging and damping (dark squares), and no cell averaging but with damping (light squares), where damping is always done by two substeps of backward Euler. As in Chapter 9, one observes that if cell averaging and damping are simultaneously applied, then a regular, second-order convergence behaviour is attained. Without damping, convergence is absent again. Without cell averaging, perhaps surprisingly, the errors are always close to those obtained with cell averaging, only their behaviour as a function of m is less regular.

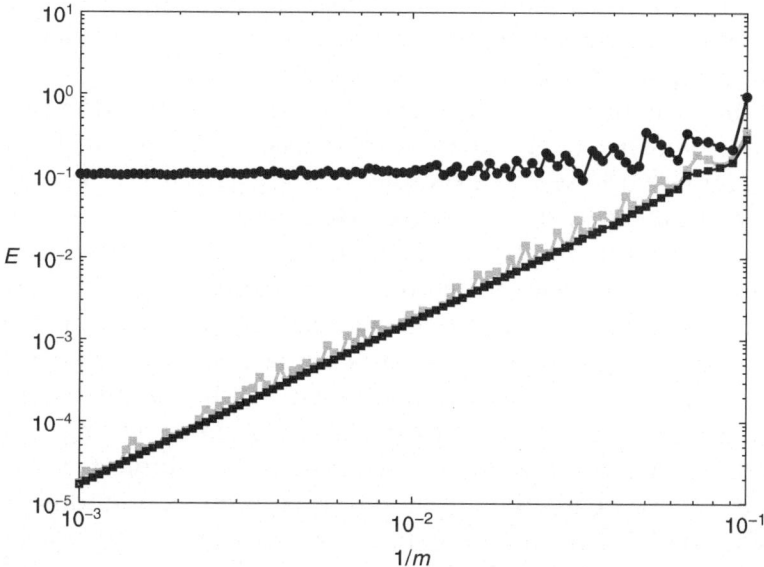

Figure 10.2 Down-and-out put option with parameter set (10.1). Total error $E^{ROI}(\Delta t;m)$ versus $1/m$ with $N = \lceil m/5 \rceil$ for $10 \leq m \leq 1000$. Crank–Nicolson method. Cell averaging without damping (dark bullets). Cell averaging with damping using two substeps (dark squares). No cell averaging but with damping using two substeps (light squares)

The down-and-out put is one of many types of barrier options traded in practice. A *down-and-out call* is the same except that it gives the holder the right to buy. An *up-and-out put (call)* gives the holder the right to sell (buy) the underlying asset for strike K at maturity T provided that the asset price does not rise above a prescribed barrier H during the lifetime $[0, T]$ of the option; otherwise the option is worthless. The numerical PDE valuation of these three barrier options proceeds similarly to that of the down-and-out put by modifying the spatial domain, the Dirichlet boundary conditions, and the initial function. Here, for accuracy, it is advantageous to have a large fraction of spatial grid points near the barrier.

A *down-and-in put* gives the holder the right to sell the underlying asset for strike K at maturity T provided that the asset price does fall below a prescribed barrier H during the lifetime $[0, T]$ of the option; otherwise the option is worthless. If the barrier is reached, one speaks in this case of a *knock-in* event. The sum of the fair values of corresponding down-and-out and down-and-in put options is identical to that of the corresponding standard put option. By using this parity relation, the down-and-in put can be numerically valued. The same goes for the "in" counterpart of any other "out" barrier option.

For all eight barrier options considered above (up/down, in/out, call/put) semi-analytical valuation formulas are known in the Black–Scholes framework, see for example [47, 90]. This is not true anymore for their so-called *discrete* variants, where the barrier constraint is enforced only a finite number of times. In this context, the options that were discussed so far are referred to as *continuous* barrier options.

We consider here a *discrete down-and-out put*. This option gives the holder the right to sell the underlying asset for strike K at maturity T provided the asset price does not fall below a barrier H at a prescribed finite, ordered set of *monitoring times* $\{\tau_1, \tau_2, \ldots, \tau_k\} \subset [0, T]$; otherwise the option is worthless. For its numerical valuation there are several changes compared to that of the continuous counterpart. The spatial domain reverts to $(0, 3K)$ and, at the lower boundary, $u(0, t) = e^{-rt}K$ if $0 \leq t < T - \tau_k$ and $u(0, t) = 0$ if $T - \tau_k \leq t \leq T$. The nonuniform spatial grid is now concentrated at the barrier instead of the strike. In Example 4.2.1 the value K is thus replaced by H. We note that for spatial accuracy it is beneficial if the barrier is located

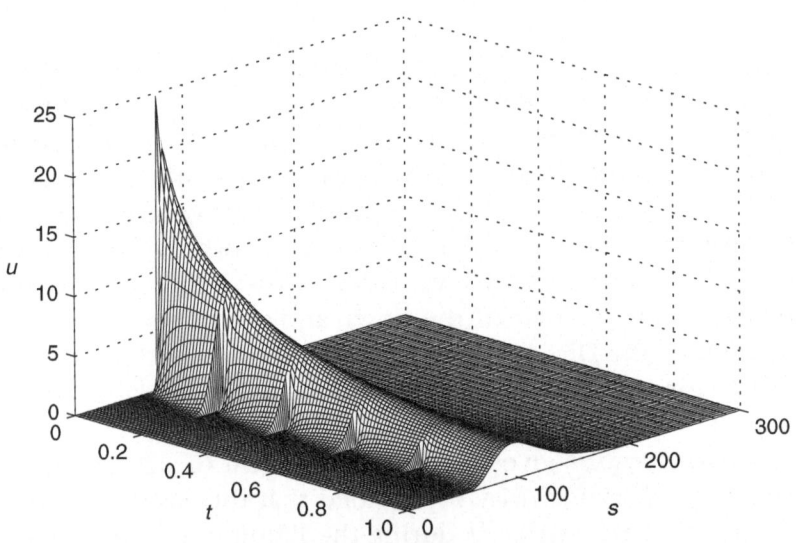

Figure 10.3 Numerically approximated discrete down-and-out put option value function on $[0, 3K] \times [0, T]$ with parameter set (10.1) and monitoring times $\tau_j = j\,T/5$ $(1 \leq j \leq 5)$

midway between two successive grid points. Cell averaging is applied to smooth the initial data at the strike, and also at the barrier if $\tau_k = T$. The time stepping proceeds as usual except that the times-till-maturity $\alpha_j = T - \tau_j$ $(1 \leq j \leq k)$ must all be temporal grid points and, for each j, once the approximation to the vector $U(\alpha_j)$ has been computed, all components corresponding to the spatial grid points $s_i \leq H$ are set equal to zero. This introduces a discontinuity at the barrier and, relatedly, damping with backward Euler is applied directly after each such event. As an illustration, Figure 10.3 displays the numerically obtained graph of the discrete down-and-out put option value function for parameter set (10.1) with five monitoring times $\tau_j = jT/5$ $(1 \leq j \leq 5)$. The discontinuities at the barrier corresponding to the monitoring times are clearly observable.

Besides the barrier options considered here, there exist many other types, for instance with double barriers, time-dependent barriers or barriers that apply during a specific time window. In principle all these options can be valued in an efficient manner through the numerical solution of an appropriate initial-boundary value problem for a PDE.

11

American-Style Options

11.1 American-Style Options

American-style options can be exercised by the holder at any given time up to and including maturity time T. This is in contrast to European-style options that were considered up to now and can only be exercised at T. Clearly, the holder of an American-style option is faced with the decision as to when it is optimal to exercise.

American options are extensively traded in practice. The fair value of any given American option is always greater than or equal to that of its European counterpart. Under the Black–Scholes framework, it can be shown that for an American call it is actually optimal to exercise at maturity and, consequently, its fair value equals that of a European call. As it turns out, however, this is an exceptional case. In general, the fair values of American options cannot be expressed in (semi-)closed analytical form. Accordingly, they are valued through numerical approximation.

We consider here the example of an *American put*. This option gives the holder the right to sell the underlying asset for strike price K at any given time up to and including maturity time T. A (semi-)closed analytical formula for the fair value of the American put is unknown. Let $u(s, t)$ denote this fair value at time $\tau = T - t$ if at that time the asset price equals s.

It is conceivable that, for any given t, there exists a value $s^*(t)$ such that if $s < s^*(t)$ it is optimal to exercise an American put and if

© The Author(s) 2017
K. in 't Hout, *Numerical Partial Differential Equations in Finance Explained*,
Financial Engineering Explained, DOI 10.1057/978-1-137-43569-9_11

$s > s^*(t)$ it is optimal to keep the option. Indeed, it can be proved that a function $s^* : [0, T] \rightarrow [0, K]$ with this property exists. Its graph is called the *early exercise boundary* or *optimal exercise boundary* or *free boundary*. For this boundary a (semi-)closed analytical formula is also unknown. At the early exercise boundary, the option value function u suffers from a lack of smoothness: it is once, but not twice, continuously differentiable there.

Let $\phi(s) = \max(K - s, 0)$ be the familiar payoff for a put option and write

$$\mathcal{A}u(s, t) = \tfrac{1}{2}\sigma^2 s^2 \frac{\partial^2 u}{\partial s^2}(s, t) + rs\frac{\partial u}{\partial s}(s, t) - ru(s, t).$$

It has been proved that under the Black–Scholes framework the following three conditions must be met at each point (s, t) with $s > 0$, $0 < t \leq T$:

$$u(s, t) \geq \phi(s), \tag{11.1a}$$

$$\frac{\partial u}{\partial t}(s, t) \geq \mathcal{A}u(s, t), \tag{11.1b}$$

$$(u(s, t) - \phi(s))\left(\frac{\partial u}{\partial t}(s, t) - \mathcal{A}u(s, t)\right) = 0. \tag{11.1c}$$

Condition (11.1a) states that the fair value of an American put is always at least equal to its instantaneous payoff. Condition (11.1b) is a so-called *partial differential inequality*. Condition (11.1c) simply means that for each given (s, t) either (11.1a) or (11.1b) must be an equality. The three conditions (11.1) together form what is called a *partial differential complementarity problem (PDCP)*. An initial condition is provided as usual by the payoff function ϕ, and at $s = 0$ one has the Dirichlet boundary condition

$$u(0, t) = K \quad \text{for } 0 \leq t \leq T. \tag{11.2}$$

Notice that a discount factor, which arises in the European case, is absent here.

The conditions (11.1) naturally induce a decomposition of the (s, t)-space: the *early exercise region* is the set of all points (s, t) where the equality $u = \phi$ holds (and the option is exercised); the *continuation region* is the set of all points (s, t) where the equality $\partial u / \partial t = \mathcal{A}u$ holds (and the option is kept). The joint boundary of these two regions is the early exercise boundary.

Assume matrix A and vector g are such that $U'(t) = AU(t) + g$ stands for any given semidiscretization of the Black–Scholes PDE on a truncated domain $(0, S_{max})$ with at $s = 0$ the Dirichlet condition (11.2) and at $s = S_{max}$ either a homogeneous Dirichlet or Neumann condition or the linear boundary condition; compare Chapter 4. Then a semidiscretization of the PDCP (11.1) on this domain and with these boundary conditions reads

$$U(t) \geq U_0, \tag{11.3a}$$

$$U'(t) \geq AU(t) + g, \tag{11.3b}$$

$$(U(t) - U_0)^{\mathrm{T}} \left(U'(t) - AU(t) - g \right) = 0 \tag{11.3c}$$

for $0 < t \leq T$ with $U(0) = U_0$ determined by the payoff function. Here inequalities for vectors are to be understood componentwise. Next, a full discretization based on the θ-method is defined by

$$U_n \geq U_0, \tag{11.4a}$$

$$(I - \theta \Delta t A) U_n \geq (I + (1 - \theta) \Delta t A) U_{n-1} + \Delta t g, \tag{11.4b}$$

$$(U_n - U_0)^{\mathrm{T}} ((I - \theta \Delta t A) U_n - (I + (1 - \theta) \Delta t A) U_{n-1} - \Delta t g) = 0 \tag{11.4c}$$

for $n = 1, 2, \ldots, N$. The three conditions (11.4) constitute a so-called *linear complementarity problem (LCP)*. This requires a new step in the numerical solution process. Despite their name, LCPs are actually *nonlinear* and they are nontrivial to solve. It is known that if all principal minors of the matrix $I - \theta \Delta t A$ are strictly positive, then (11.4) always possesses a unique solution. Matrices with this property are called P-matrices.

11.2   LCP Solution Methods

Solution methods for LCPs have been widely studied in the literature. We discuss here three approximation approaches that are often used for the application under consideration. Each of these successively generates for $n = 1, 2, \ldots, N$ approximations $\widehat{U}_n$ to the vectors U_n defined by (11.4).

The *explicit payoff method* for (11.4) is the most basic approach and yields

$$(I - \theta \Delta t A)\bar{U}_n = (I + (1 - \theta)\Delta t A)\widehat{U}_{n-1} + \Delta t\, g, \qquad (11.5a)$$

$$\widehat{U}_n = \max\{\bar{U}_n\,, U_0\}. \qquad (11.5b)$$

Here $\widehat{U}_0 = U_0$ and the maximum of two vectors is to be understood componentwise. Method (11.5) can be viewed as first performing a time step by ignoring the American constraint, and next applying this constraint explicitly. The computational cost per time step is essentially the same as that in the case of the European counterpart of the option, which is very favourable. The obtained accuracy with the explicit payoff method is often relatively low, however.

The *Ikonen-Toivanen (IT) splitting method* for (11.4) is a more advanced approach and yields

$$(I - \theta \Delta t A)\bar{U}_n = (I + (1 - \theta)\Delta t A)\widehat{U}_{n-1} + \Delta t\, g + \Delta t\, \widehat{\lambda}_{n-1}, \quad (11.6a)$$

$$\begin{cases} \widehat{U}_n - \bar{U}_n - \Delta t\,(\widehat{\lambda}_n - \widehat{\lambda}_{n-1}) = 0, \\[2mm] \widehat{U}_n \geq U_0, \quad \widehat{\lambda}_n \geq 0, \quad (\widehat{U}_n - U_0)^{\mathrm{T}} \widehat{\lambda}_n = 0, \end{cases} \qquad (11.6b)$$

with $\widehat{\lambda}_0 = 0$. The vector $\widehat{U}_n$ together with the auxiliary vector $\widehat{\lambda}_n$ are computed in two stages. In the first stage an intermediate approximation $\bar{U}_n$ is defined by the system of linear equations (11.6a). In the second stage, $\bar{U}_n$ and $\widehat{\lambda}_{n-1}$ are updated to $\widehat{U}_n$ and $\widehat{\lambda}_n$ by (11.6b). It is readily verified that these updates are given by the simple, explicit formulas

$$\widehat{U}_n = \max\left\{\bar{U}_n - \Delta t\,\widehat{\lambda}_{n-1}\,, U_0\right\},$$

$$\widehat{\lambda}_n = \max\left\{0\,, \widehat{\lambda}_{n-1} + (U_0 - \bar{U}_n)/\Delta t\right\}.$$

As for the explicit payoff method, the computational cost per time step with the IT splitting method is essentially the same as that in the case of the European counterpart of the option.

Let $G > 0$ be any given fixed large number. The *penalty method* for (11.4) reads

$$\left(I - \theta \Delta t A + P_n^{(k)}\right) \bar{U}_n^{(k+1)} = (I + (1-\theta)\Delta t A)\widehat{U}_{n-1} + \Delta t\, g + P_n^{(k)} U_0$$

for $k = 0, 1, \ldots, l-1$ and $\widehat{U}_n = \bar{U}_n^{(l)}$. $\qquad$ (11.7)

Hence, this yields an iteration in each time step. The initial iterate $\bar{U}_n^{(0)} = \widehat{U}_{n-1}$. Next, $P_n^{(k)}$ for $0 \le k < l$ is defined as the diagonal matrix where the i-th diagonal entry is equal to G whenever $\bar{U}_{n,i}^{(k)} < U_{0,i}$ and zero otherwise. The $P_n^{(k)}$ can be viewed as penalty matrices forcing $\widehat{U}_n$ to approximately satisfy the LCP (11.4). If the penalty factor G is taken sufficiently large, this approximation is expected to be highly accurate. In each time step of the penalty method, $l \ge 1$ linear systems need to be solved, involving different matrices. Accordingly, the penalty method is computationally more expensive per time step than the two methods (11.5), (11.6). The number of iterations l can either be chosen a priori or be defined a posteriori by a convergence criterion. A natural criterion is

$$\max_i \frac{|\bar{U}_{n,i}^{(k+1)} - \bar{U}_{n,i}^{(k)}|}{\max\{1, |\bar{U}_{n,i}^{(k+1)}|\}} < tol \quad \text{or} \quad P_n^{(k+1)} = P_n^{(k)}, \qquad (11.8)$$

with a given sufficiently small tolerance $tol > 0$. Let ε denote the machine precision of the computer. A rule of thumb for the choice of penalty factor is then

$$G \approx \alpha \frac{tol}{\varepsilon} \quad \text{with } \alpha = 10^{-2}. \qquad (11.9)$$

11.3 Numerical Study

We consider an American put option with

$$K = 100, \ T = 0.5, \ r = 0.02, \ \sigma = 0.25. \tag{11.10}$$

Figure 11.1 shows the numerically approximated graph of the option value function (dark curve) together with the graph of the payoff function (light curve) on $[\frac{1}{2}K, \frac{3}{2}K]$ for $t = T$. Clearly, the option value is always greater than or equal to the corresponding payoff value. At the point $s = s^*(T)$ on the left of K where the two graphs meet, their derivatives with respect to s are equal. This is known as *smooth pasting*. Figure 11.2 displays the numerically approximated early exercise boundary. It is directly obtained by verifying whether or not $\widehat{U}_{n,i} = U_{0,i}$ holds. Notice that the function s^* which defines the early exercise boundary varies strongly near $t = 0$, that is, when actual time is close to maturity. We mention that $s^*(T) \approx 73.4$.

For the experiments, as in foregoing chapters, the spatial domain is truncated to $(0, 3K)$ and semidiscretization is performed on the nonuniform grid from Example 4.2.1 with second-order central formulas for convection and diffusion, using formula B for convection. Cell averaging is applied to smooth the initial data at the strike. The only change with respect to the semidiscretization for the European put option case is a different Dirichlet boundary condition at $s = 0$, see (11.2). The matrix A thus remains the same, whereas the vector $g(t)$ changes slightly and becomes independent of t.

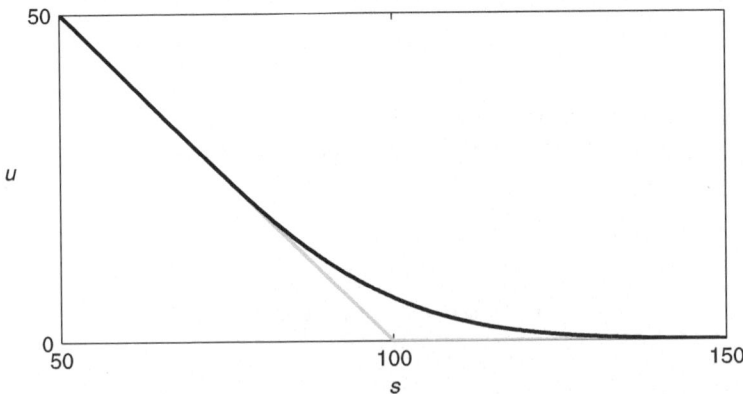

Figure 11.1   Dark: numerically approximated American put option value function on $[\frac{1}{2}K, \frac{3}{2}K]$ for $t = T$ and parameter set (11.10). Light: payoff function

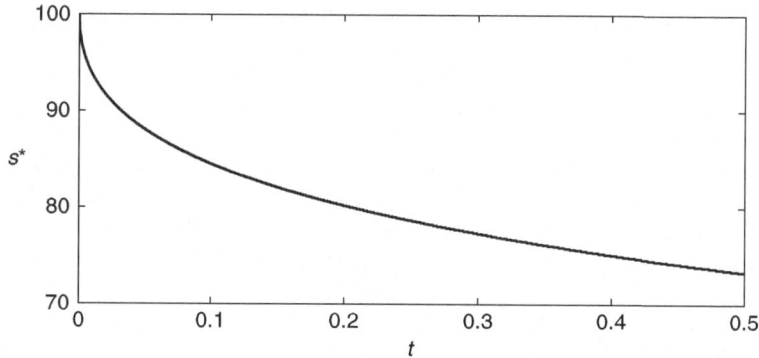

Figure 11.2   Numerically approximated early exercise boundary for the American put option and parameter set (11.10)

To gain insight into the different convergence behaviours of the methods (11.5), (11.6), (11.7) we examine the norm of the *temporal discretization error* on a region of interest defined by

$$\widehat{e}^{ROI}(\Delta t;m) = \max\{|U_i(T) - \widehat{U}_{N,i}|: 0 \le i \le m,\ \tfrac{4}{5}K < s_i < \tfrac{5}{4}K\},$$

where $U(T)$ denotes the exact solution[1] to the semidiscrete PDCP (11.3) for $t = T$. For each of the three methods the first time step is always replaced by two substeps of the same method using half the step size and $\theta = 1$, akin to backward Euler damping. For the penalty method the convergence criterion (11.8) is used and parameter values $tol = 10^{-8}$, $G = 10^6$ are chosen. Notice that the early exercise point $s^*(T)$ lies outside the region of interest, so the focus is on a region where the exact solution u is smooth. We further emphasize that the spatial discretization error is not incorporated here.

Consider $N = \lceil m/2 \rceil$ and $10 \le m \le 1000$. Figure 11.3 displays the temporal discretization errors for the explicit payoff method (bullets), the IT splitting method (squares) and the penalty method (triangles) both for $\theta = \tfrac{1}{2}$ (dark) and $\theta = 1$ (light). For $\theta = 1$ the errors obtained with the three methods are very close to each other. They show a first-order convergence behaviour, as might be expected. For $\theta = \tfrac{1}{2}$ the errors are substantially smaller than for $\theta = 1$. In this case the

---

[1] Recall that when computing discretization errors the semidiscrete solution is approximated to a high accuracy by applying a suitable temporal discretization method using a very small step size.

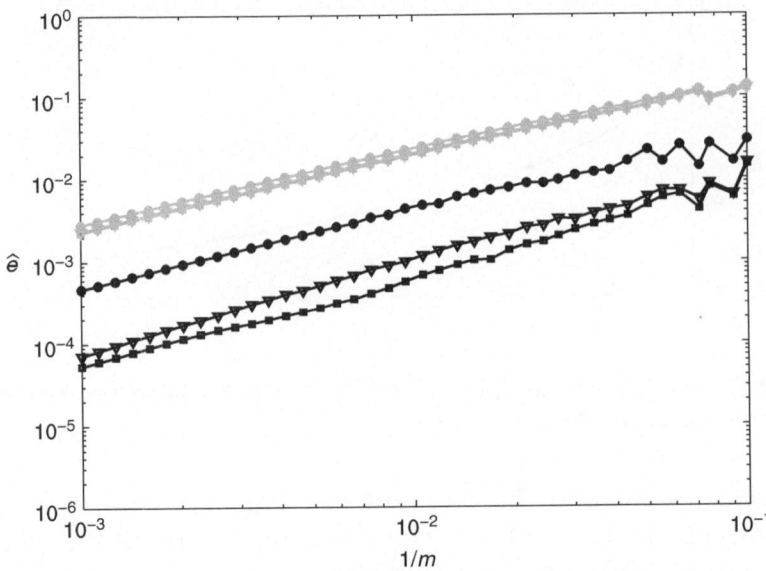

Figure 11.3 American put option with parameter set (11.10). Temporal error $\hat{e}^{ROI}(\Delta t;m)$ versus $1/m$ with $N = \lceil m/2 \rceil$ for $10 \leq m \leq 1000$. Constant step sizes. Backward Euler: light. Crank–Nicolson: dark. Method (11.5): bullets. Method (11.6): squares. Method (11.7): triangles

explicit payoff method is the least accurate. The errors for the IT splitting method and penalty method are relatively close to each other, where the former is slightly more accurate. However, if also $\theta = \frac{1}{2}$ an approximate first-order temporal convergence behaviour is observed.

It can be argued that *variable step sizes* are more natural than constant step sizes, in view of the strongly varying early exercise boundary near $t = 0$. All temporal discretization methods under consideration in this book are adapted straightforwardly to variable step sizes, in particular the methods (11.5), (11.6) and (11.7). Consider temporal grid points defined by

$$t_n = \left(\frac{n}{N}\right)^2 T \quad \text{for } n = 0, 1, 2, \ldots, N. \tag{11.11}$$

The corresponding step sizes are smallest near $t = 0$ and grow linearly with n. Figure 11.4 displays the temporal discretization errors for the same situation as that of Figure 11.3, except that now the temporal grid points (11.11) are used. For $\theta = 1$ the obtained errors are

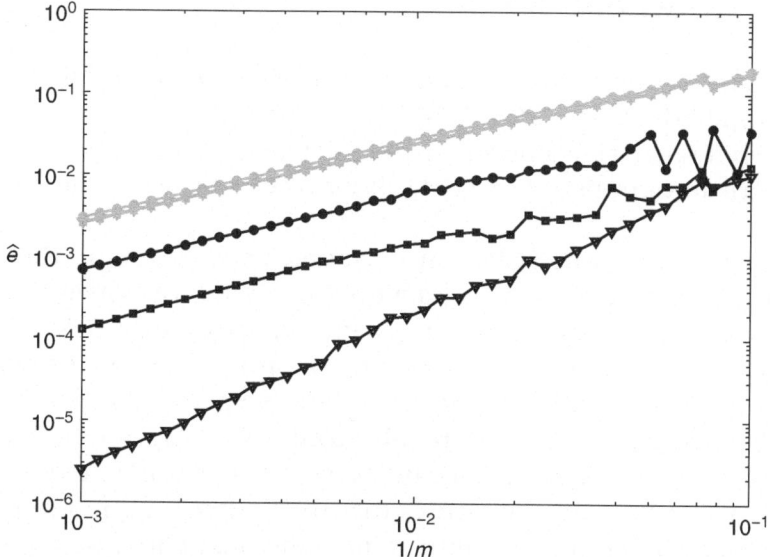

Figure 11.4 American put option with parameter set (11.10). Temporal error $\widehat{e}^{ROI}(\Delta t; m)$ versus $1/m$ with $N = \lceil m/2 \rceil$ for $10 \leq m \leq 1000$. Variable step sizes. Backward Euler: light. Crank–Nicolson: dark. Method (11.5): bullets. Method (11.6): squares. Method (11.7): triangles

generally about the same as in the constant step size case. For the penalty method with $\theta = \frac{1}{2}$ there is a substantial improvement, however. This method now reveals a second-order temporal convergence behaviour, which is as desired. Furthermore, the number of iterations l per time step turns out to be 1.3 on average, which is very favourable. For the explicit payoff method and the IT splitting method with $\theta = \frac{1}{2}$ the errors are in general larger than in the case of constant step sizes, and the observed temporal convergence order remains close to one.

Of the methods under consideration, the IT splitting method with $\theta = \frac{1}{2}$ for constant step sizes and the penalty method with $\theta = \frac{1}{2}$ for variable step sizes are preferable in this experiment. The former method only shows a first-order temporal convergence behaviour, but the linear system in each time step can be solved very efficiently by using an upfront LU factorization as before. The latter method reveals a second-order temporal convergence behaviour, but solving the linear system(s) in each time step is computationally more expensive. The actual choice of method depends among others on the accuracy that is required.

11.4   Notes and References

American-style options and their numerical valuation are discussed in for example the texts [21, 47, 58, 79, 85, 90, 94].

First references considering the PDCP formulation for the fair values of American options are [53, 91]. Prior to this, it was indirectly used in [6].

Many properties of LCPs can be found in for example the books [4, 14]. There exists a wide range of methods for solving LCPs in general and for applications in finance in particular. The classical Brennan–Schwartz algorithm [6, 53] is optimal in the present example, as it solves the LCP in each time step exactly, but it is not clear how to extend it to more advanced American option valuation problems. In this chapter, three modern approximation methods have been considered with the virtues that they are natural in the present time-dependent context, simple to implement and broadly applicable. The penalty method was proposed for American option valuation in [25, 95] and the IT splitting method in [50, 51]. The choice (11.9) for the penalty factor is recommended by [24] and incorporates floating point round-off considerations [46].

For the penalty method various convergence results relevant to financial applications are available in the literature; compare [25, 45]. For the IT splitting method the theory is under development, see [30] for a recent convergence result. Their application with variable step sizes, in particular corresponding to (11.11), is discussed in for example [25, 51, 73].

12

Merton Model

12.1 Merton Model

It is a well-known phenomenon in financial markets that sudden, large movements in asset prices occur every now and then. This can happen for example after major news events. Already in 1976, Merton [63] proposed to add a jump term to the geometric Brownian motion in order to obtain a better model for the asset price evolution. The jumps are assumed to follow a compound Poisson process, so that they arrive randomly according to a Poisson process and their size is random as well compare for example [80]. When a jump occurs, the price of the asset is modelled by multiplying its price at the time instant just before the jump with a given positive random variable Y. Thus a fall in the asset price corresponds to $Y \in (0, 1)$ and a rise to $Y > 1$. In this chapter we consider Y to be lognormally distributed. It then has the probability density function

$$f(y) = \frac{1}{y\delta\sqrt{2\pi}} \exp\left(-\frac{(\ln y - \gamma)^2}{2\delta^2}\right) \quad (y > 0), \qquad (12.1)$$

where γ and δ denote given real constants with $\delta > 0$ that are equal to the mean and standard deviation, respectively, of the normal random variable $\ln(Y)$. The pertinent stochastic process for the asset price is called the *Merton model* and forms a particular instance of a general jump-diffusion model.

© The Author(s) 2017
K. in 't Hout, *Numerical Partial Differential Equations in Finance Explained*,
Financial Engineering Explained, DOI 10.1057/978-1-137-43569-9_12

Denote by λ the intensity of the Poisson process and let κ be the expected relative jump size,

$$\kappa = \mathbb{E}[Y-1] = \int_0^\infty (y-1)f(y)dy = \exp\left(\gamma + \tfrac{1}{2}\delta^2\right) - 1.$$

Under certain assumptions on the market and the hedging approach, it can be argued that if the geometric Brownian motion (1.2) for the underlying asset price is replaced by the Merton model, then the fair value $u(s, t)$ of a European-style option satisfies the equation

$$\frac{\partial u}{\partial t}(s, t) = \tfrac{1}{2}\sigma^2 s^2 \frac{\partial^2 u}{\partial s^2}(s, t) + r_0 s \frac{\partial u}{\partial s}(s, t) - r_1 u(s, t) + \lambda \int_0^\infty u(sy, t)f(y)dy$$

$$(12.2)$$

for $s > 0$ and $0 < t \leq T$ with $r_0 = r - \lambda\kappa$ and $r_1 = r + \lambda$. Comparing this to the Black–Scholes PDE (1.3) one observes that a new term has appeared, which is an integral involving the option value function u. Accordingly, (12.2) is called a *partial integro-differential equation (PIDE)*. The integral term is *nonlocal* - as opposed to the derivative terms, which are *local*. It involves, for any given s and t, the values of the option for all possible underlying asset prices. An adequate numerical treatment of this integral term is crucial in order to arrive at an efficient numerical solution method for (12.2).

In the following we shall consider as an example the numerical valuation of a European put option under the Merton model. The PIDE (12.2) is complemented with the same initial condition and Dirichlet boundary condition at $s = 0$ as in the case of the Black–Scholes PDE. The option value function u is known in semi-closed analytical form [63]:

$$u(s, t) = \sum_{k=0}^{\infty} \frac{(\mu t)^k}{k!} e^{-\mu t} u_k(s, t) \qquad (12.3)$$

whenever $s > 0, 0 < t \leq T$. Here $\mu = \lambda(1+\kappa)$ and $u_k(s, t)$, for any given integer $k \geq 0$, denotes the Black–Scholes value (1.6b) for a put option upon replacing r and σ^2 by, respectively,

$$r - \lambda\kappa + \frac{k\ln(1+\kappa)}{t} \quad \text{and} \quad \sigma^2 + \frac{k\delta^2}{t}.$$

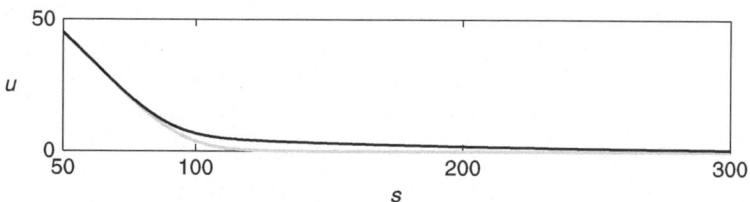

Figure 12.1 Exact put option value functions on $[\frac{1}{2}K, 3K]$ for $t = T$ and parameter set (12.4). Black–Scholes value: light. Merton value: dark

For the numerical experiments we take the financial parameter set

$$K = 100, \ T = 1, \ r = 0.05, \ \sigma = 0.15, \ \lambda = 0.1, \ \gamma = -0.9, \ \delta = 0.45.$$
$$(12.4)$$

Figure 12.1 displays the corresponding graphs of the put option value function under the geometric Brownian motion (Black–Scholes) model and the Merton model. The figure illustrates in particular that the option value under the Merton model is nonnegligible on a much larger region of asset prices s than that under the Black–Scholes model, as one would expect.

12.2 Spatial Discretization

As the starting point for the semidiscretization of (12.2) the spatial domain is truncated to $(0, 5K)$, the familiar nonuniform grid from Example 4.2.1 is considered, second-order central formulas for convection and diffusion are applied with formula B for convection, and cell averaging is performed. At $S_{\max} = 5K$ a homogeneous Dirichlet condition is taken.

As previously, denote the semidiscrete approximation to $u(s_i, t)$ by $U_i(t)$ for $1 \leq i \leq m - 1$. Set $U_0(t) = e^{-rt}K$ and $U_m(t) = 0$ in view of the Dirichlet boundary conditions.

To complete the semidiscretization of (12.2) a discretization of the integral term is required. Let $s = s_i$ with arbitrary $1 \leq i \leq m-1$ be given and define $f_i(x) = f(x/s_i)/s_i$ for $x > 0$ with $f(0) = 0$. Slightly rewriting the integral, yields

$$\int_0^\infty u(s_i y, t)f(y)dy = \int_0^\infty u(x, t)f_i(x)dx \approx \int_0^{S_{\max}} u(x, t)f_i(x)dx.$$

It is now convenient to have an approximation to $u(x, t)$ available for all $x \in (0, S_{max})$, that is, not just at the discrete spatial grid, but on the whole truncated spatial domain. To this purpose an interpolation procedure can be employed. Since second-order spatial convergence is aimed for, piecewise linear interpolation is a natural candidate. Hence, for any given $1 \leq j \leq m$, consider the approximation

$$u(x, t) \approx \frac{s_j - x}{h_j} U_{j-1}(t) + \frac{x - s_{j-1}}{h_j} U_j(t) \quad (s_{j-1} \leq x \leq s_j).$$

Combining the above leads to

$$\int_0^\infty u(s_i y, t) f(y) dy \approx \sum_{j=1}^m \int_{s_{j-1}}^{s_j} \left[\frac{s_j - x}{h_j} U_{j-1}(t) + \frac{x - s_{j-1}}{h_j} U_j(t) \right] f_i(x) dx.$$

Define two integrals involving the probability density function f,

$$\mathcal{J}_{0,i,j} = \int_{s_{j-1}}^{s_j} f_i(x) dx \quad \text{and} \quad \mathcal{J}_{1,i,j} = \int_{s_{j-1}}^{s_j} x f_i(x) dx. \tag{12.5}$$

Then the following semidiscrete approximation to the integral term in (12.2) at $s = s_i$ is obtained,

$$\lambda \int_0^\infty u(s_i y, t) f(y) dy \approx \sum_{j=1}^{m-1} A_{0,i,j} U_j(t) + g_{0,i}(t), \tag{12.6}$$

where

$$A_{0,i,j} = \lambda \left[\frac{\mathcal{J}_{1,i,j} - s_{j-1} \mathcal{J}_{0,i,j}}{h_j} + \frac{s_{j+1} \mathcal{J}_{0,i,j+1} - \mathcal{J}_{1,i,j+1}}{h_{j+1}} \right], \tag{12.7a}$$

$$g_{0,i}(t) = \lambda \left[\frac{s_1 \mathcal{J}_{0,i,1} - \mathcal{J}_{1,i,1}}{h_1} U_0(t) + \frac{\mathcal{J}_{1,i,m} - s_{m-1} \mathcal{J}_{0,i,m}}{h_m} U_m(t) \right]. \tag{12.7b}$$

For f given by (12.1) the two integrals (12.5) can be expressed in terms of the standard normal cumulative distribution function $\mathcal{N}$. Define

$$\psi_{0,i}(s) = \mathcal{N} \left(\frac{\ln (s/s_i) - \gamma}{\delta} \right), \tag{12.8a}$$

$$\psi_{1,i}(s) = s_i \exp \left(\gamma + \tfrac{1}{2} \delta^2 \right) \mathcal{N} \left(\frac{\ln (s/s_i) - \gamma}{\delta} - \delta \right), \tag{12.8b}$$

for $s > 0$ and $\psi_{0,i}(0) = \psi_{1,i}(0) = 0$. Then it is readily shown that

$$\mathcal{J}_{0,i,j} = \psi_{0,i}(s_j) - \psi_{0,i}(s_{j-1}) \quad \text{and} \quad \mathcal{J}_{1,i,j} = \psi_{1,i}(s_j) - \psi_{1,i}(s_{j-1}). \quad (12.9)$$

Consider the matrix and vector

$$A_0 = (A_{0,i,j})_{i,j=1}^{m-1} \quad \text{and} \quad g_0(t) = (g_{0,i}(t))_{i=1}^{m-1}$$

and denote by A_1 and $g_1(t)$ the matrix and vector defining the semidiscretization of (12.2) without the integral term. Then the semidiscretization of the complete PIDE (12.2) is of the form (4.1) with

$$A = A_0 + A_1 \quad \text{and} \quad g(t) = g_0(t) + g_1(t).$$

12.3 IMEX Schemes

For the temporal discretization of the obtained semidiscrete PIDE (12.2) the Crank-Nicolson method (with damping) could be used. This yields the recurrence relation (7.3) for the fully discrete approximations U_n at the temporal grid points. There is an important change, however, in moving from a PDE to a PIDE problem. Due to the presence of the integral term, the matrix A has lost the sparsity property and is now a full matrix: all its entries are nonzero in general. As a consequence, solving the linear system (7.3) in each time step has turned into a computationally expensive task.

To overcome this disadvantage, an *implicit-explicit (IMEX) scheme* can be applied for the temporal discretization. An IMEX scheme effectively employs the splitting $A = A_0 + A_1$ of the semidiscrete operator A such that in each time step only the sparse matrix A_1 (the PDE part) is treated in an implicit way and the full matrix A_0 (the integral part) is treated in an explicit way. This yields a major reduction in computational cost per time step compared to standard (nonsplit) temporal discretization schemes such as the Crank-Nicolson method.

We consider here a specific IMEX scheme that can be viewed as a blend between the trapezoidal rule (the Crank-Nicolson method) and its *explicit variant*, also known as the modified Euler method or the Heun method. It generates, in a one-step manner, successive approximations U_n for $n = 1, 2, \ldots, N$ by

$$
\begin{cases}
Y_0 = U_{n-1} + \Delta t A U_{n-1} + \Delta t g(t_{n-1}), \\
\widehat{Y}_0 = Y_0 + \frac{1}{2}\Delta t A_0 \left(Y_0 - U_{n-1}\right) + \frac{1}{2}\Delta t (g_0(t_n) - g_0(t_{n-1})), \\
Y_1 = \widehat{Y}_0 + \frac{1}{2}\Delta t A_1 \left(Y_1 - U_{n-1}\right) + \frac{1}{2}\Delta t (g_1(t_n) - g_1(t_{n-1})), \\
U_n = Y_1.
\end{cases}
\tag{12.10}
$$

The vectors Y_0, $\widehat{Y}_0$ and Y_1 are internal stages that can be discarded after each time step. The first two stages are explicit, whereas the third stage is implicit. It is readily verified that by formally setting $A_0 = 0$, $g_0 = 0$ the IMEX scheme (12.10) reduces to the familiar trapezoidal rule, and that by formally setting $A_1 = 0$, $g_1 = 0$ it reduces to its explicit variant. The internal stages can be eliminated, leading to

$$
\left(I - \tfrac{1}{2}\Delta t A_1\right) U_n = \left(I + \Delta t A_0 + \tfrac{1}{2}\Delta t A_1 + \tfrac{1}{2}(\Delta t)^2 A_0 A\right) U_{n-1} +
$$
$$
\tfrac{1}{2}\Delta t (g(t_{n-1}) + \Delta t A_0 g(t_{n-1}) + g(t_n)).
\tag{12.11}
$$

Since the matrix $I - \frac{1}{2}\Delta t A_1$ is tridiagonal, this linear system can be solved very efficiently in every time step by using an upfront LU factorization, as before. Further, it is often not necessary to compute the full matrix-vector product $A_0 g(t)$ on the right-hand side in every time step. In the present example it is equal to $e^{-rt} A_0 g(0)$, so that it suffices to compute $A_0 g(0)$ once, upfront, and then perform only a scalar multiplication of this vector in every time step.

One easily shows that A_0 and g_0 are both uniformly bounded, independently of the spatial mesh. In view of this, treating the semi-discretized integral term explicitly in a temporal discretization scheme may be expected to work well.

12.4 Numerical Study

We discuss numerical experiments with the discretization of the valuation problem for the put option under the Merton model described in the foregoing two sections. Here the Crank–Nicolson method and IMEX scheme are both applied with damping, using two substeps of

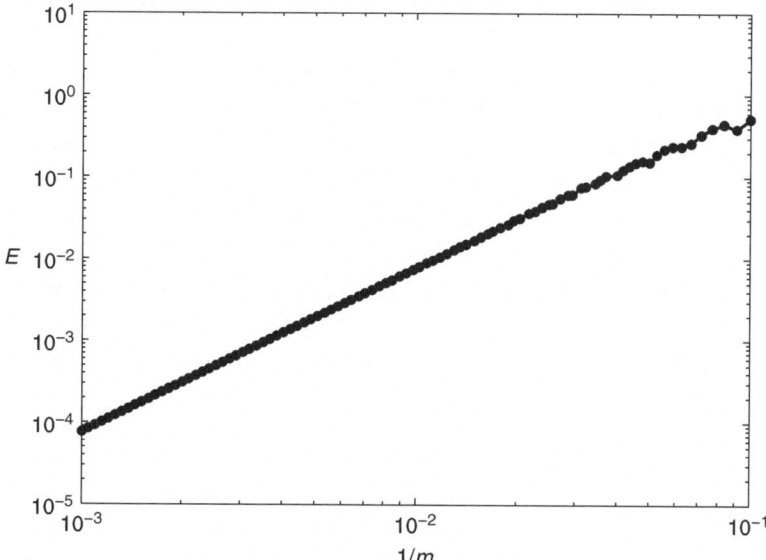

Figure 12.2 Put option under Merton model with parameter set (12.4). Total error $E^{ROI}(\Delta t;m)$ versus $1/m$ with $N = \lceil m/3 \rceil$ for $10 \leq m \leq 1000$. Schemes: Crank–Nicolson (light bullets), IMEX (dark bullets)

backward Euler. The two schemes are found to yield visually the same temporal discretization errors for any given m and N under consideration. A second-order temporal convergence behaviour is observed, uniformly in m, which agrees with their theoretical orders. Due to its lower computational cost, the IMEX scheme is preferable to the Crank–Nicolson method in terms of efficiency.

Figure 12.2 displays the norm of the total discretization error $E^{ROI}(\Delta t;m)$ versus $1/m$ with $\Delta t = T/N$, $N = \lceil m/3 \rceil$, $10 \leq m \leq 1000$ and region of interest given by $\frac{1}{2}K < s < \frac{3}{2}K$. The results for the Crank–Nicolson method are indicated by light bullets and those for the IMEX scheme by dark bullets. Only the latter are visible, however, since the two graphs lie on top of each other. The total errors are seen to be $\mathcal{O}(m^{-2})$, thus a favourable second-order convergence of the full discretization is obtained.

12.5 Notes and References

Jump-diffusion models and numerical option valuation under such models via PIDEs are discussed for example in the books [21, 79,

85, 90]. For financial modelling with general Lévy processes see for instance [12, 77].

An overview and analysis of IMEX schemes when applied to PDEs, without integral term, is given in the book [49], where in particular (12.10) is considered. The application of IMEX schemes to PIDEs in finance has been studied in for example [2, 8, 13, 23, 56, 75, 76].

Various authors have considered a different set-up, whereby the Fast Fourier Transform (FFT) can be employed for the efficient computation of matrix-vector products with the pertinent full matrix A_0. This is a natural idea because the integral term itself can be regarded, after a simple change of variables, as a convolution. For details on this approach, see for example [1, 33].

13

Two-Asset Options

13.1 Two-Asset Options

Multi-asset options depend on more than one underlying asset. In this chapter we shall focus on two-asset options. Assuming the Black–Scholes framework, the price evolution of two assets is given by two geometric Brownian motions that may be correlated to each other. The fair value of a European-style option is then a function of three independent real variables: $u = u(s_1, s_2, t)$ where s_i represents the price of asset i at time $T - t$ for $i = 1, 2$. Financial option valuation theory yields that the function u satisfies the PDE

$$\frac{\partial u}{\partial t} = \tfrac{1}{2}\sigma_1^2 s_1^2 \frac{\partial^2 u}{\partial s_1^2} + \rho\sigma_1\sigma_2 s_1 s_2 \frac{\partial^2 u}{\partial s_1 \partial s_2} + \tfrac{1}{2}\sigma_2^2 s_2^2 \frac{\partial^2 u}{\partial s_2^2} + rs_1 \frac{\partial u}{\partial s_1} + rs_2 \frac{\partial u}{\partial s_2} - ru$$

(13.1)

for $s_1 > 0$, $s_2 > 0$ and $0 < t \leq T$. Here the positive constant σ_i denotes the volatility of the price of asset i for $i = 1, 2$ and the constant $\rho \in [-1, 1]$ stands for the correlation factor pertinent to the two underlying asset price processes. The constant r denotes the risk-free interest rate as before.

The PDE (13.1) is of the time-dependent convection-diffusion-reaction kind, like the Black–Scholes PDE (1.3). However, besides the time variable t, there are now two independent variables, namely s_1 and s_2. In view of this, the PDE (13.1) is said to be *two-dimensional*. Next, a *mixed derivative term* $u_{s_1 s_2}$ is present whenever ρ is nonzero.

© The Author(s) 2017
K. in 't Hout, *Numerical Partial Differential Equations in Finance Explained*,
Financial Engineering Explained, DOI 10.1057/978-1-137-43569-9_13

Mixed derivative terms are ubiquitous in multi-dimensional financial option valuation PDEs due to the fact that the underlying stochastic processes are almost always correlated. This is distinctive from many other application areas, where such terms do not arise often.

As a concrete example for numerical experiments we consider in this chapter a *European call option on the maximum of two assets*. It has the payoff

$$\phi(s_1, s_2) = \max\left(\max(s_1, s_2) - K, 0\right) \quad \text{for } s_1 \geq 0, s_2 \geq 0. \quad (13.2)$$

The fair value function for this option is known in semi-closed analytical form and is provided in Appendix D. Figure 13.1 displays the payoff function ϕ and Figure 13.2 the option value function u on the (s_1, s_2)-domain $[0, 3K] \times [0, 3K]$ for $t = T$ and the sample parameter set

$$K = 100, \; T = 0.75, \; r = 0.02, \; \sigma_1 = 0.30, \; \sigma_2 = 0.50, \; \rho = 0.40. \quad (13.3)$$

As usual, the payoff function (13.2) defines the initial condition for (13.1), that is, $u(s_1, s_2, 0) = \phi(s_1, s_2)$.

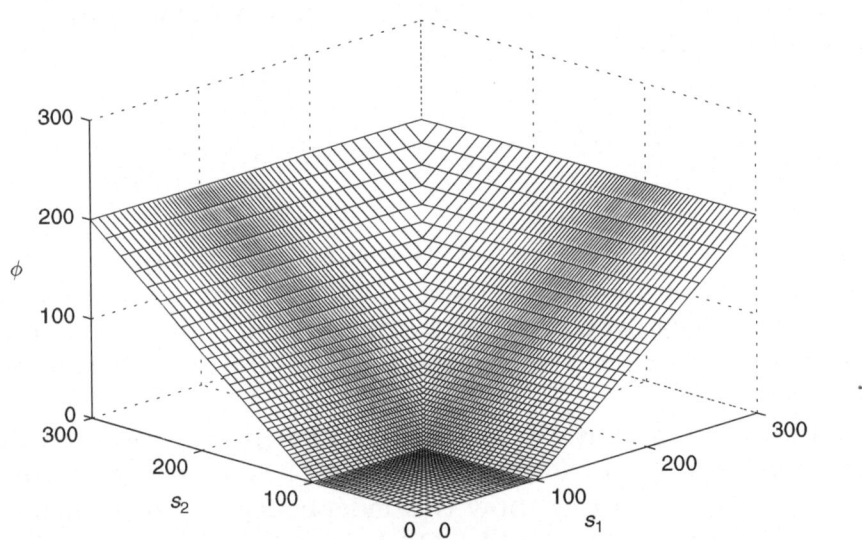

Figure 13.1   Payoff function for call on the maximum of two assets option on $[0, 3K] \times [0, 3K]$ if $K = 100$

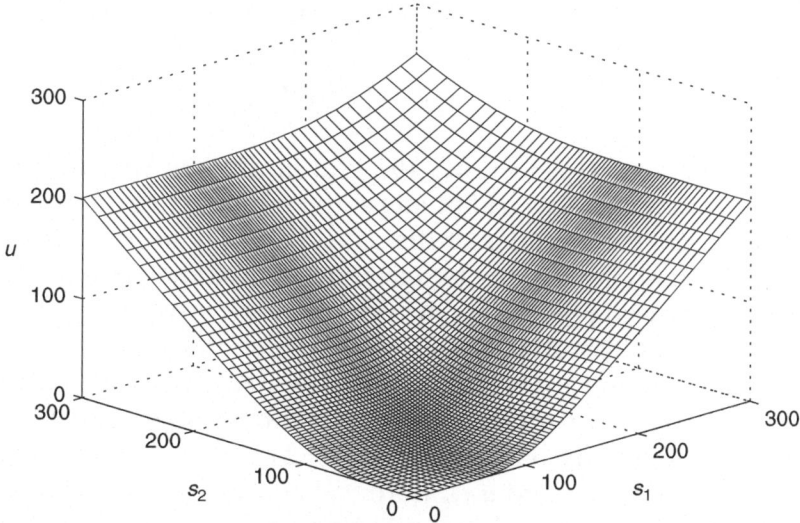

Figure 13.2   Exact option value function for call on the maximum of two assets on $[0, 3K] \times [0, 3K]$ for $t = T$ and parameter set (13.3)

13.2 Spatial Discretization

13.2.1 Finite Difference Discretization

For the spatial discretization of the PDE (13.1) the spatial domain is truncated to the square $(0, S_{\max}) \times (0, S_{\max})$ with $S_{\max} = 5K$. Boundary conditions need to be prescribed at its four sides. At the two sides satisfying $s_1 = 0$ and $s_2 = 0$, respectively, it can be shown that (13.1) is itself fulfilled. Thus one obtains the original, one-dimensional Black–Scholes PDE there, with volatilities $\sigma = \sigma_2$ and $\sigma = \sigma_1$, respectively. This forms a new type of boundary condition. Using the Black–Scholes formula from Chapter 1, one could also impose Dirichlet conditions at $s_1 = 0$ and $s_2 = 0$, but it is easier and more generic to refrain from doing this. At the two "far" sides, satisfying $s_1 = S_{\max}$ and $s_2 = S_{\max}$, linear boundary conditions are selected, that is,

$$u_{s_1 s_1}(S_{\max}, s_2, t) = 0 \quad (0 < s_2 < S_{\max}, \, 0 \le t \le T),$$
$$u_{s_2 s_2}(s_1, S_{\max}, t) = 0 \quad (0 < s_1 < S_{\max}, \, 0 \le t \le T).$$

For the spatial grid the Cartesian product of two nonuniform grids given by Example 4.2.1 (with $S_{\min}=0$ and $L=K/3$) is taken. Accordingly, the set of spatial grid points at which u is approximated becomes

$$(s_{1,i}, s_{2,j}) \quad \text{for } 0 \le i \le m_1, \, 0 \le j \le m_2, \tag{13.4}$$

where $s_{k,0}, s_{k,1}, \ldots, s_{k,m_k}$ denotes the grid in the k-th spatial direction ($k = 1, 2$). Notice that the grid points at the boundary of the square are all included since the values of the exact solution are not prescribed there. By construction, the two-dimensional spatial grid possesses relatively many grid points near the important location $(s_1, s_2) = (K, K)$.

For the spatial discretization second-order central formulas for convection and diffusion are taken to semidiscretize the spatial derivatives $u_{s_1}, u_{s_2}, u_{s_1 s_1}, u_{s_2 s_2}$. At the boundary of the truncated spatial domain we have the following. Firstly, (for $k = 1, 2$) if $s_k = 0$, then the parts in (13.1) involving u_{s_k} and $u_{s_k s_k}$ both vanish. As a consequence, they are trivially dealt with in the semidiscretization. Next, (for $k = 1, 2$) if $s_k = S_{\max}$, then the derivative $u_{s_k s_k}$ vanishes by the linear boundary condition and the derivative u_{s_k} is semidiscretized by means of the first-order backward formula as in Section 4.1.

It remains for us to consider semidiscretization of the mixed derivative $u_{s_1 s_2}$. A useful approach to deriving finite difference formulas for this term is to regard it as successive differentiation in the s_1- and s_2-directions and then numerically mimicking this. If $\omega_{k,i,-1}$, $\omega_{k,i,0}$, $\omega_{k,i,1}$ denote the coefficients of any given finite difference formula for convection in the s_k-direction at the point $s_{k,i}$, then the following *finite difference approximation for the mixed derivative* is obtained,

$$u_{s_1 s_2}(s_{1,i}, s_{2,j}, t) \approx \sum_{p=-1}^{1} \sum_{q=-1}^{1} \omega_{1,i,p} \, \omega_{2,j,q} \, u(s_{1,i+p}, s_{2,j+q}, t). \tag{13.5}$$

With central formula A for convection in both directions, this approximation uses four function values of u, centred about the point $(s_{1,i}, s_{2,j})$. With central formula B in both directions, it uses nine function values in general. If the spatial grid were uniform in both directions, with constant mesh widths h_1 and h_2, respectively, then the above approximation reduces with either formula to

$$\frac{u(s_{1,i+1}, s_{2,j+1}, t) + u(s_{1,i-1}, s_{2,j-1}, t) - u(s_{1,i-1}, s_{2,j+1}, t) - u(s_{1,i+1}, s_{2,j-1}, t)}{4h_1 h_2}.$$

On smooth nonuniform Cartesian grids, as considered here, it can be shown by Taylor's theorem that the truncation error with the approximation (13.5) using either formula A or B is $\mathcal{O}((\Delta\xi)^2 + \Delta\xi\,\Delta\eta + (\Delta\eta)^2)$ whenever u is smooth, where $\Delta\xi$ and $\Delta\eta$ denote the constant mesh widths of the underlying artificial grids for the s_1- and s_2-directions, respectively; compare Section 4.2.

Concerning the boundary of the spatial domain, the part in the PDE (13.1) involving the mixed derivative term vanishes if $s_1 = 0$ or $s_2 = 0$. If $s_1 = S_{\max}$ or $s_2 = S_{\max}$, then the finite difference treatment of the mixed derivative term is naturally induced by that of the two convection terms at these boundary sides through (13.5).

13.2.2 Cell Averaging

As in the one-dimensional case, it is beneficial also in the multi-dimensional case to slightly modify the initial data at the spatial grid by applying cell averaging near the points of nonsmoothness of the payoff function. For the payoff (13.2), these points make up three line segments, compare Figure 13.1:

$$s_1 = K \quad \text{and} \quad 0 \le s_2 < K;$$
$$s_2 = K \quad \text{and} \quad 0 \le s_1 < K;$$
$$s_2 = s_1 \quad \text{and} \quad K \le s_1 \le S_{\max}.$$

Define for $k = 1, 2$

$$s_{k,l+1/2} = \tfrac{1}{2}(s_{k,l} + s_{k,l+1}) \qquad \text{for } 0 \le l < m_k,$$
$$h_{k,l+1/2} = s_{k,l+1/2} - s_{k,l-1/2} \quad \text{for } 0 \le l \le m_k,$$

with $s_{k,-1/2} = 0$ and $s_{k,m_k+1/2} = S_{\max}$ at the boundary of the spatial domain. Consider any given cell

$$[s_{1,i-1/2}, s_{1,i+1/2}) \times [s_{2,j-1/2}, s_{2,j+1/2}).$$

If this has a nonempty intersection with the set of points where the payoff ϕ is nonsmooth, then the pointwise value $\phi(s_{1,i}, s_{2,j})$ in the initial data is replaced by the cell averaged value

$$\frac{1}{h_{1,i+1/2}\, h_{2,j+1/2}} \int_{s_{1,i-1/2}}^{s_{1,i+1/2}} \int_{s_{2,j-1/2}}^{s_{2,j+1/2}} \phi(s_1, s_2)\, ds_2\, ds_1. \tag{13.6}$$

For many payoff functions, including (13.2), this integral is readily calculated. Otherwise a numerical integration rule can be applied for its approximation.

13.2.3 Semidiscrete System

The semidiscretization outlined above of the initial-boundary value problem for the PDE (13.1) leads to an initial value problem for a system of ODEs. This system can be written in the form (4.1) with size $v = (m_1+1)(m_2+1)$. Here the vector $U(t)$ represents the semidiscrete approximation to the exact solution $u(\cdot, \cdot, t)$ on the Cartesian grid (13.4), where we choose the lexicographical order. The vector $g(t)$ is identically equal to zero since there are no contributions from the boundary conditions; this somewhat simplifies the implementation. Thus

$$U'(t) = AU(t) \quad (0 < t \le T).$$

The matrix A can be conveniently expressed by means of the *Kronecker product*, denoted by $\otimes$. If P and $Q = (q_{ij})$ are any given two matrices, then by definition

$$Q \otimes P = \left(q_{ij}P\right).$$

Hence, each entry in Q is replaced by that entry multiplied with the matrix P. This obviously yields a (much) larger matrix. Note that the Kronecker product is a noncommutative operation. For the matrix A, it can be shown that

$$
\begin{align}
A &= A_0 + A_1 + A_2, \tag{13.7a}\\
A_0 &= \rho\, \sigma_1 \sigma_2 \left(X_2 D_2^{(1)}\right) \otimes \left(X_1 D_1^{(1)}\right), \tag{13.7b}\\
A_1 &= I_2 \otimes \left(\tfrac{1}{2}\sigma_1^2 X_1^2 D_1^{(2)} + rX_1 D_1^{(1)} - \tfrac{1}{2}rI_1\right), \tag{13.7c}\\
A_2 &= \left(\tfrac{1}{2}\sigma_2^2 X_2^2 D_2^{(2)} + rX_2 D_2^{(1)} - \tfrac{1}{2}rI_2\right) \otimes I_1. \tag{13.7d}
\end{align}
$$

Here A_0 represents the mixed derivative part in (13.1) and (for $k = 1, 2$) A_k represents all derivative terms in the k-th spatial direction plus half of the $-ru$ term. The $D_k^{(l)}, X_k, I_k$ are given $(m_k+1) \times (m_k+1)$ matrices for $k, l \in \{1, 2\}$ where I_k is the identity matrix, X_k is the diagonal matrix

$$X_k = \text{diag}(s_{k,0}, s_{k,1}, \ldots, s_{k,m_k})$$

and $D_k^{(l)}$ denotes the matrix representing numerical differentiation of order l in the k-th spatial direction by the relevant finite difference formula. More precisely, if $1 \le i \le m_k - 1$, then the i-th row of $D_k^{(l)}$ has as entries in columns $i - 1, i, i + 1$ the pertinent finite difference coefficients $\omega_{k,i,-1}, \omega_{k,i,0}, \omega_{k,i,1}$, respectively, and is zero elsewhere. The entries in the top row, indicated by number 0, are not important as they are multiplied by zero in (13.7). One can therefore set this row equal to zero. In view of the treatment of the linear boundary condition at $s_k = S_{\max}$, the last row of $D_k^{(1)}$, indicated by number m_k, corresponds to the first-order backward formula for convection and the last row of $D_k^{(2)}$ is equal to zero.

The numerical differentiation matrices $D_k^{(l)}$ are all tridiagonal. It follows that the matrix A_1 is tridiagonal too and A_2 is essentially tridiagonal, that is, up to a simple permutation. The matrix A, on the other

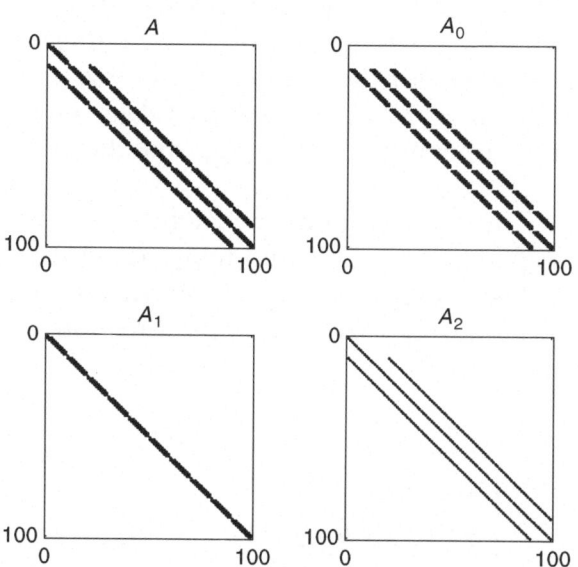

Figure 13.3 Sparsity pattern of A, A_0, A_1, A_2 given by (13.7) if $m_1 = m_2 = 9$. Finite difference approximation (13.5) of mixed derivative based on formula B

hand, is not tridiagonal. Besides a main tridiagonal band, it possesses a tridiagonal band on either side that lies "far away", at a distance directly proportional to m_1. This particular structure is a consequence of the two-dimensionality of the PDE (13.1). As an illustration Figure 13.3 displays the sparsity pattern (all nonzero entries) of each of the four matrices A, A_0, A_1, A_2 for a small-sized sample case.

13.3   ADI Schemes

For the temporal discretization of the semidiscretized two-dimensional PDE (13.1) the Crank–Nicolson method with damping can be applied where the obtained linear system in each time step is solved by means of straightforward LU factorization. Due to the size and structure of the present matrix A, discussed above, the matrices L and U exhibit substantial *fill-in*, however. Both possess many more nonzero entries than the linear system matrix $I - \frac{1}{2}\Delta t A$ itself. This results in a much larger amount of floating-point operations per time step than if A were tridiagonal. For higher-dimensional option valuation PDEs, such as for arbitrary multi-asset options, this increase in computational cost is even more pronounced. One way around this is to use a sparse direct solver or a suitable iterative solver for the linear systems. An alternative approach, which we discuss here, is to apply an *Alternating Direction Implicit (ADI) scheme* for the temporal discretization. ADI schemes effectively employ the splitting $A = A_0 + A_1 + A_2$ given by (13.7). These schemes are of the operator splitting type, like the IMEX scheme from Chapter 12.

Let $\theta \in (0, 1]$ be a given parameter, let $N \geq 1$ be any given integer and let step size $\Delta t = T/N$ and temporal grid points $t_n = n\Delta t$. The *Douglas scheme* constitutes the basic ADI scheme. It successively defines, in a one-step fashion, approximations U_n to $U(t_n)$ for $n = 1, 2, \ldots, N$ by

$$
\begin{cases}
Y_0 = U_{n-1} + \Delta t A U_{n-1}, \\
Y_1 = Y_0 + \theta \Delta t A_1 (Y_1 - U_{n-1}), \\
Y_2 = Y_1 + \theta \Delta t A_2 (Y_2 - U_{n-1}), \\
U_n = Y_2.
\end{cases}
\tag{13.8}
$$

The Y_k's denote internal stages that are discarded after each time step. The scheme (13.8) is implicit and bears resemblance to the θ-method. The computational cost per time step of the Douglas scheme is often much lower, however, than that of the θ-method. The Douglas scheme starts with an explicit stage, defining Y_0, which is followed by two implicit unidirectional stages, defining Y_1 and Y_2. The latter vectors are the solutions to two linear systems with matrices $I - \theta \Delta t A_1$ and $I - \theta \Delta t A_2$, respectively. Since these matrices are both (essentially) tridiagonal, the corresponding linear systems *can* be solved very efficiently by means of LU factorization. The number of floating-point operations per time step of the Douglas scheme is then directly proportional to the size ν of A, that is, the number of spatial grid points, which is optimal.

An important useful feature of the Douglas scheme is that it treats the mixed derivative part in an *explicit* way: the matrix A_0 occurs only in the first stage. Hence, it is not required to solve linear systems involving this matrix. There is a disadvantage to (13.8), however, since the first stage is just a forward Euler step. Consequently, its temporal convergence order does not exceed one whenever A_0 is nonzero, which is the common situation in practice. The following two extensions achieve temporal convergence order equal to 2 also if A_0 is nonzero.

The first extension is the *Modified Craig–Sneyd (MCS) scheme*:

$$
\begin{cases}
Y_0 = U_{n-1} + \Delta t A U_{n-1}, \\
Y_1 = Y_0 + \theta \Delta t A_1 (Y_1 - U_{n-1}), \\
Y_2 = Y_1 + \theta \Delta t A_2 (Y_2 - U_{n-1}), \\
\widehat{Y}_0 = Y_0 + \theta \Delta t A_0 (Y_2 - U_{n-1}), \\
\widetilde{Y}_0 = \widehat{Y}_0 + \left(\tfrac{1}{2} - \theta \right) \Delta t A (Y_2 - U_{n-1}), \\
\widetilde{Y}_1 = \widetilde{Y}_0 + \theta \Delta t A_1 (\widetilde{Y}_1 - U_{n-1}), \\
\widetilde{Y}_2 = \widetilde{Y}_1 + \theta \Delta t A_2 (\widetilde{Y}_2 - U_{n-1}), \\
U_n = \widetilde{Y}_2.
\end{cases}
\tag{13.9}
$$

The second extension is the *Hundsdorfer-Verwer (HV) scheme*:

$$
\begin{cases}
Y_0 = U_{n-1} + \Delta t A U_{n-1}, \\
Y_1 = Y_0 + \theta \Delta t A_1 (Y_1 - U_{n-1}), \\
Y_2 = Y_1 + \theta \Delta t A_2 (Y_2 - U_{n-1}), \\
\widetilde{Y}_0 = Y_0 + \tfrac{1}{2} \Delta t A (Y_2 - U_{n-1}), \\
\widetilde{Y}_1 = \widetilde{Y}_0 + \theta \Delta t A_1 (\widetilde{Y}_1 - Y_2), \\
\widetilde{Y}_2 = \widetilde{Y}_1 + \theta \Delta t A_2 (\widetilde{Y}_2 - Y_2), \\
U_n = \widetilde{Y}_2.
\end{cases}
\qquad (13.10)
$$

The initial three stages of the MCS and HV schemes are the same as those of the Douglas scheme. These schemes then perform a subsequent explicit stage, which is followed again by two implicit unidirectional stages. The special case $\theta = \tfrac{1}{2}$ of (13.9) is called the *Craig-Sneyd (CS) scheme*. Like the Douglas scheme, the MCS and HV schemes also treat the mixed derivative part in an explicit manner. The amount of computational work per time step for the two schemes (13.9) and (13.10) is about twice that of the basic scheme (13.8).

For the MCS and HV schemes it has been proved that, under stability and smoothness conditions, the temporal convergence bound (7.4) holds in the scaled Euclidean norm with order $q = 2$ and constant $\widehat{C}$ independent of the spatial mesh widths. The pertinent conditions admit arbitrary (nonzero) correlation factors ρ. For this generic situation, a recommended lower bound on the parameter θ for the MCS scheme is $\theta \geq \tfrac{1}{3}$ and for the HV scheme it is given by $\theta \geq 1 - \tfrac{1}{2}\sqrt{2} \approx 0.293$ (if the PDE is diffusion-dominated) and $\theta \geq \tfrac{1}{2} + \tfrac{1}{6}\sqrt{3} \approx 0.789$ (if the PDE is convection-dominated). For the HV scheme the first of these two bounds is satisfactory in the case of the PDE (13.1) unless one of the volatilities σ_1, σ_2 is very low. Practical experience with the above ADI schemes shows that a smaller value θ often yields a better (smaller) error constant $\widehat{C}$.

13.4 Numerical Study

We consider experiments with the discretization of the valuation problem for the European call option on the maximum of two assets described in the foregoing two sections. The parameter set (13.3) is chosen. For the spatial discretization, the same number of grid points in the two directions is taken, $m_1 = m_2 = m$, and formula B is used for the finite difference discretization of the convection term as well as for the mixed derivative term through (13.5). For the temporal discretization we select the Crank–Nicolson scheme together with four ADI schemes:

- Douglas with $\theta = \frac{1}{2}$,
- MCS with $\theta = \frac{1}{3}$,
- MCS with $\theta = \frac{1}{2}$ (CS),
- HV with $\theta = 1 - \frac{1}{2}\sqrt{2}$.

Damping is always applied, using two substeps of backward Euler at the start.

To study the convergence behaviour of the different temporal discretization schemes we consider the norm of the temporal discretization error on a natural region of interest,

$$\widehat{e}^{ROI}(\Delta t; m) = \max\{|U_\kappa(T) - U_{N,\kappa}|: 0 \le i,j \le m, \tfrac{1}{2}K < s_{1,i}, s_{2,j} < \tfrac{3}{2}K\},$$

where $\kappa = \kappa(i,j)$ denotes the index such that $U_\kappa(T)$ and $U_{N,\kappa}$ correspond to the spatial grid point $(s_{1,i}, s_{2,j})$. Figure 13.4 shows the errors $\widehat{e}^{ROI}(\Delta t; m)$ versus $1/m$ with $N = m$ for $10 \le m \le 100$. For the Douglas scheme (light squares) a first-order temporal convergence behaviour is found. This convergence order is due to the presence of a mixed derivative part (ρ is nonzero), as discussed in the previous section. For the other four schemes a favourable second-order temporal convergence behaviour is observed. The Crank–Nicolson scheme (light bullets) yields the most accurate results, but is computationally more expensive. The MCS scheme with $\theta = \frac{1}{3}$ (dark squares) and the HV scheme with $\theta = 1 - \frac{1}{2}\sqrt{2}$ (dark triangles) both achieve almost the same accuracy as the Crank–Nicolson scheme, but at a much lower computational cost in general. The temporal errors for the CS scheme (dark bullets) are approximately twice as large as those for the MCS and HV schemes, for about the same amount of computational work.

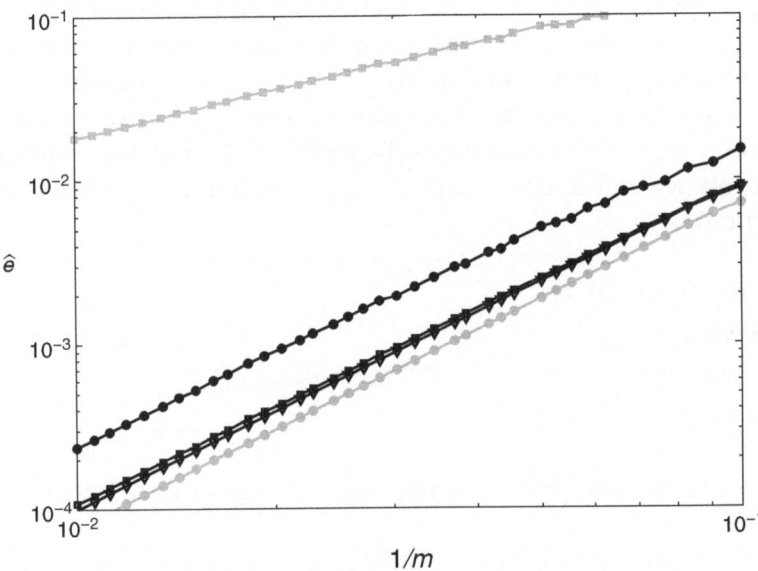

Figure 13.4   Call option on the maximum of two assets with parameter set (13.3). Temporal error $\widehat{e}^{ROI}(\Delta t;m)$ versus $1/m$ with $N = m$ for $10 \leq m \leq 100$. Schemes: Crank–Nicolson (light bullets), Douglas with $\theta = \frac{1}{2}$ (light squares), CS (dark bullets), MCS with $\theta = \frac{1}{3}$ (dark squares), HV with $\theta = 1 - \frac{1}{2}\sqrt{2}$ (dark triangles)

We next examine the norm of the total discretization error,

$$E^{ROI}(\Delta t;m) = \max\{|u_{i,j} - U_{N,\kappa}|: 0 \leq i,j \leq m,\ \tfrac{1}{2}K < s_{1,i},\ s_{2,j} < \tfrac{3}{2}K\},$$

where $u_{i,j} = u(s_{1,i}, s_{2,j}, T)$. Figure 13.5 displays the total errors versus $1/m$ with $N = m$ and $10 \leq m \leq 100$. For the Douglas scheme, if $m \gtrsim 25$, then the total errors are only $\mathcal{O}(m^{-1})$. This is a consequence of the first-order temporal convergence. For the other four schemes the total errors are essentially the same and $\mathcal{O}(m^{-2})$. Hence, second-order convergence is attained with these schemes. Taking into account the amount of computational work, the CS, MCS and HV schemes are clearly preferable over the Crank–Nicolson scheme.

It is finally mentioned that if the finite difference formula A is applied instead of formula B, then the obtained total errors with Crank–Nicolson and the latter ADI schemes are always somewhat larger (at most a factor 2), but a second-order convergence behaviour is retained.

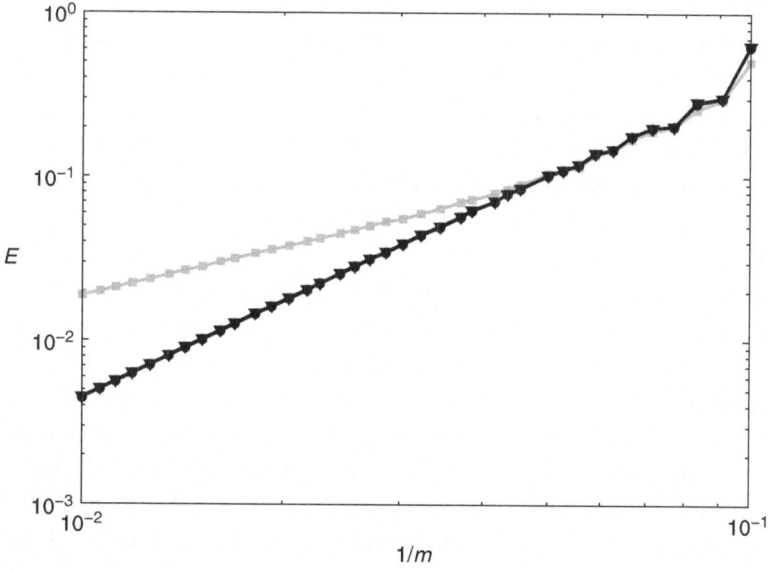

Figure 13.5   Call option on the maximum of two assets with parameter set (13.3). Total error $E^{ROI}(\Delta t; m)$ versus $1/m$ with $N = m$ for $10 \leq m \leq 100$. Schemes: Crank–Nicolson (light bullets), Douglas with $\theta = \frac{1}{2}$ (light squares), CS (dark bullets), MCS with $\theta = \frac{1}{3}$ (dark squares), HV with $\theta = 1 - \frac{1}{2}\sqrt{2}$ (dark triangles)

13.5   Notes and References

Multi-asset options and their numerical valuation are discussed in for example the books [11, 21, 58, 79, 85, 90, 94].

ADI schemes were introduced in 1955 by Peaceman and Rachford [69] and Douglas [18] and subsequently developed by Douglas and Rachford [20], Brian [7] and Douglas [19] for solving the two- and three-dimensional heat equations. These original ADI discretizations can all be written in the form (13.8) with $A_0 = 0$ and with an additional internal stage Y_3 in the case of three dimensions.

In [60, 61] the ADI idea was applied to two-dimensional convection-diffusion equations containing a mixed derivative term, where this term was dealt with in an explicit way. The pertinent scheme can be formulated as (13.8) with $\theta = \frac{1}{2}$.

In [16] a novel ADI scheme was proposed for arbitrary multi-dimensional pure diffusion equations with mixed derivative terms, where these terms are also handled explicitly. In the case of two dimensions it becomes the scheme (13.9) with $\theta = \frac{1}{2}$.

The general scheme (13.9) was introduced in [42] and the scheme (13.10) in [48, 49], both also for arbitrary spatial dimensions. The application of the latter scheme to PDEs involving mixed derivative terms was first analyzed in [41].

ADI schemes have been employed in computational finance since the mid-1990s, compare [58, 85, 90]. For actual applications of ADI schemes to advanced contemporary financial PDEs, see for example [17, 28, 29, 35, 52].

An ample stability and convergence analysis for ADI schemes applied to PDEs without mixed derivative terms is presented in [49]. At present the most comprehensive stability results for ADI schemes that are relevant to PDEs with mixed derivative terms are provided in [36, 37, 41, 42, 64]. These references contain in particular the recommended lower bounds on θ from Section 13.3. The temporal convergence result stated in the same section has been established in [43, 44]. A detailed convergence analysis for the scheme (13.9) in the case of nonsmooth initial data has been presented in [93].

For more information on operator splitting methods in finance, see also [38].

Appendix A: Wiener Process

A *standard Brownian motion* or *Wiener process* in $\mathbb{R}$ is a family of random variables W_τ ($\tau \geq 0$) on a same probability space $(\Omega, \mathcal{F}, \mathbb{P})$ such that the following four conditions hold:

(i) $W_0(\omega) = 0$ for almost all $\omega \in \Omega$,

(ii) for every $\tau, \upsilon \geq 0$ the random variable $W_{\tau+\upsilon} - W_\tau$ is normally distributed with mean 0 and variance υ,

(iii) for every $0 = \tau_0 < \tau_1 < \cdots < \tau_N$ the random variables $W_{\tau_n} - W_{\tau_{n-1}}$ ($n = 1, 2, \ldots, N$) are independent,

(iv) for almost every $\omega \in \Omega$ the sample path $W_\tau(\omega)$ ($\tau \geq 0$) is continuous.

As an illustration, five sample paths on $[0, 1]$ are displayed in the next figure.

© The Author(s) 2017 **113**
K. in 't Hout, *Numerical Partial Differential Equations in Finance Explained*,
Financial Engineering Explained, DOI 10.1057/978-1-137-43569-9

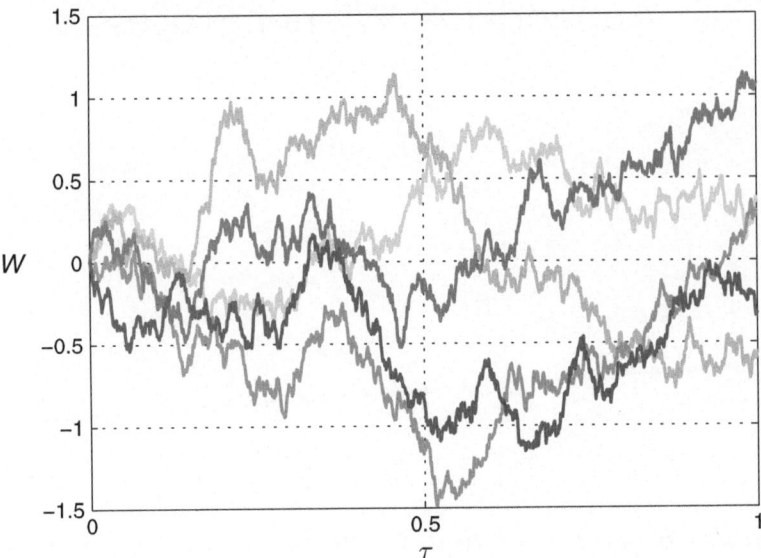

Appendix B: Feynman–Kac Theorem

Let $T > 0$ and $r \geq 0$ be given constants and let α, β, ϕ be given functions from $\mathbb{R}$ into $\mathbb{R}$. For any given $t \in (0, T]$ consider the Itô stochastic differential equation

$$dS_\tau = \alpha(S_\tau)\, d\tau + \beta(S_\tau)\, dW_\tau \quad (T - t < \tau \leq T)$$

and for any given $s \in \mathbb{R}$ define

$$u(s, t) = \mathbb{E}[\, e^{-rt}\, \phi(S_T) \,|\, S_{T-t} = s\,].$$

Then under appropriate assumptions on α, β, ϕ the function u is well-defined and is the unique solution to the PDE

$$\frac{\partial u}{\partial t}(s, t) = \tfrac{1}{2}\beta(s)^2 \frac{\partial^2 u}{\partial s^2}(s, t) + \alpha(s)\frac{\partial u}{\partial s}(s, t) - ru(s, t) \qquad \text{(B.1)}$$

for $s \in \mathbb{R}$, $0 < t \leq T$ with initial condition $u(s, 0) = \phi(s)$ for $s \in \mathbb{R}$.

Full details and a proof of the Feynman-Kac theorem are given in for example [59, 68, 80]. Various useful variants and extensions are also discussed there.

(B.1) is often referred to as a Feynman–Kac equation. It is further closely related to a so-called backward Kolmogorov equation.

© The Author(s) 2017 **115**
K. in 't Hout, *Numerical Partial Differential Equations in Finance Explained*,
Financial Engineering Explained, DOI 10.1057/978-1-137-43569-9

Appendix C: Down-and-Out Put Option Value

Let the barrier $H < K$. Then for $s \geq H$, $0 < t \leq T$ the fair value $u(s,t)$ of a down-and-out put option under the Black–Scholes framework is given by:

$$u(s,t) = s\left[\mathcal{N}(d_1) - \mathcal{N}(d_3)\right] - e^{-rt}K\left[\mathcal{N}(d_2) - \mathcal{N}(d_4)\right] +$$

$$s(H/s)^{2\lambda}\left[\mathcal{N}(d_5) - \mathcal{N}(d_7)\right] - e^{-rt}K(H/s)^{2\lambda-2}\left[\mathcal{N}(d_6) - \mathcal{N}(d_8)\right],$$

where

$$\lambda = \frac{r}{\sigma^2} + \frac{1}{2}, \qquad\qquad \mu = \sigma\sqrt{t},$$

$$d_1 = \frac{\ln(s/K)}{\mu} + \lambda\mu, \qquad d_2 = d_1 - \mu,$$

$$d_3 = \frac{\ln(s/H)}{\mu} + \lambda\mu, \qquad d_4 = d_3 - \mu,$$

$$d_5 = \frac{\ln(H/s)}{\mu} + \lambda\mu, \qquad d_6 = d_5 - \mu,$$

$$d_7 = \frac{\ln(H^2/(sK))}{\mu} + \lambda\mu, \; d_8 = d_7 - \mu,$$

see for example [47].

© The Author(s) 2017
K. in 't Hout, *Numerical Partial Differential Equations in Finance Explained*,
Financial Engineering Explained, DOI 10.1057/978-1-137-43569-9

Appendix D: Max-of-Two-Assets Call Option Value

For $s_1 > 0$, $s_2 > 0$ and $0 < t \leq T$ the fair value $u(s_1, s_2, t)$ of a call option on the maximum of two assets under the Black–Scholes framework is given by [84]:

$$u(s_1, s_2, t) = s_1 \, \mathcal{N}(d_1, d, \rho_1) + s_2 \, \mathcal{N}(d_2, -d + \sigma\sqrt{t}, \rho_2) -$$

$$e^{-rt} K \left[1 - \mathcal{N}(-d_1 + \sigma_1\sqrt{t}, -d_2 + \sigma_2\sqrt{t}, \rho) \right],$$

where

$$\sigma = \sqrt{\sigma_1^2 + \sigma_2^2 - 2\rho\sigma_1\sigma_2},$$

$$\rho_1 = \frac{\sigma_1 - \rho\sigma_2}{\sigma},$$

$$\rho_2 = \frac{\sigma_2 - \rho\sigma_1}{\sigma},$$

$$d = \frac{\ln(s_1/s_2) + \frac{1}{2}\sigma^2 t}{\sigma\sqrt{t}},$$

$$d_i = \frac{\ln(s_i/K) + (r + \frac{1}{2}\sigma_i^2)t}{\sigma_i\sqrt{t}} \quad (i = 1, 2),$$

and $\mathcal{N}(\cdot, \cdot, \rho)$ denotes the bivariate normal cumulative distribution function with zero mean and covariance matrix

$$\begin{pmatrix} 1 & \rho \\ \rho & 1 \end{pmatrix}.$$

© The Author(s) 2017
K. in 't Hout, *Numerical Partial Differential Equations in Finance Explained*,
Financial Engineering Explained, DOI 10.1057/978-1-137-43569-9

Bibliography

[1] A. Almendral & C. W. Oosterlee, *Numerical valuation of options with jumps in the underlying*, Appl. Numer. Math. **53** (2005) 1–18.

[2] L. Andersen & J. Andreasen, *Jump-diffusion processes: volatility smile fitting and numerical methods for option pricing*, Rev. Deriv. Res. **4** (2000) 231–262.

[3] L. Bachelier, *Théorie de la spéculation*, Annales Scientifiques de l'École Normale Supérieure **3** (1900) 21–86.

[4] A. Berman & R. J. Plemmons, Nonnegative Matrices in the Mathematical Sciences, SIAM, 1994.

[5] F. Black & M. Scholes, *The pricing of options and corporate liabilities*, J. Polit. Econ. **81** (1973) 637–654.

[6] M. J. Brennan & E. S. Schwartz, *The valuation of American put options*, J. Finan. **32** (1977) 449–462.

[7] P. L. T. Brian, *A finite-difference method of high-order accuracy for the solution of three-dimensional transient heat conduction problems*, AIChE J. **7** (1961) 367–370.

[8] M. Briani, R. Natalini & G. Russo, *Implicit-explicit numerical schemes for jump-diffusion processes*, Calcolo **44** (2007) 33–57.

[9] J. C. Butcher, Numerical Methods for Ordinary Differential Equations, 2nd ed., Wiley, 2008.

[10] L. Capriotti, Y. Jiang & A. Macrina, *Real-time risk management: an AAD-PDE approach*, Int. J. Finan. Eng. **2** (2015) 1550039.

[11] I. J. Clark, Foreign Exchange Option Pricing, Wiley, 2011.

[12] R. Cont & P. Tankov, Financial Modelling with Jump Processes, Chapman & Hall, 2003.

[13] R. Cont & E. Voltchkova, *A finite difference scheme for option pricing in jump diffusion and exponential Lévy models*, SIAM J. Numer. Anal. **43** (2005) 1596–1626.

[14] R. W. Cottle, J. S. Pang & R. E. Stone, The Linear Complementarity Problem, Academic Press, 1992.

© The Author(s) 2017
K. in 't Hout, *Numerical Partial Differential Equations in Finance Explained*,
Financial Engineering Explained, DOI 10.1057/978-1-137-43569-9

[15] R. Courant, K. O. Friedrichs & H. Lewy, *Über die partiellen Differenzengleichungen der mathematischen Physik*, Math. Anal. **100** (1928) 32-74.

[16] I. J. D. Craig & A. D. Sneyd, *An alternating-direction implicit scheme for parabolic equations with mixed derivatives*, Comp. Math. Appl. **16** (1988) 341-350.

[17] D. M. Dang, C. C. Christara, K. R. Jackson & A. Lakhany, *An efficient numerical partial differential equation approach for pricing foreign exchange interest rate hybrid derivatives*, J. Comp. Finan. **18** (2015) 39-93.

[18] J. Douglas, *On the numerical integration of $u_{xx} + u_{yy} = u_t$ by implicit methods*, J. Soc. Ind. Appl. Math. **3** (1955) 42-65.

[19] J. Douglas, *Alternating direction methods for three space variables*, Numer. Math. **4** (1962) 41-63.

[20] J. Douglas & H. H. Rachford, *On the numerical solution of heat conduction problems in two and three space variables*, Trans. Amer. Math. Soc. **82** (1956) 421-439.

[21] D. J. Duffy, Finite Difference Methods in Financial Engineering, Wiley, 2006.

[22] L. C. Evans, Partial Differential Equations, 2nd ed., AMS, 2010.

[23] L. Feng & V. Linetsky, *Pricing options in jump-diffusion models: an extrapolation approach*, Oper. Res. **56** (2008) 304-325.

[24] P. A. Forsyth, private communication, 2016.

[25] P. A. Forsyth & K. R. Vetzal, *Quadratic convergence for valuing American options using a penalty method*, SIAM J. Sci. Comp. **23** (2002) 2095-2122.

[26] M. B. Giles, *Adjoint methods for option pricing, Greeks and calibration using PDEs and SDEs*, Lecture notes, Oxford Univ., 2012.

[27] M. B. Giles & R. Carter, *Convergence analysis of Crank-Nicolson and Rannacher time-marching*, J. Comp. Finan. **9** (2006) 89-112.

[28] T. Haentjens, *Efficient and stable numerical solution of the Heston-Cox-Ingersoll-Ross partial differential equation by alternating direction implicit finite difference schemes*, Int. J. Comp. Math. **90** (2013) 2409-2430.

[29] T. Haentjens & K. J. in 't Hout, *Alternating direction implicit finite difference schemes for the Heston-Hull-White partial differential equation*, J. Comp. Finan. **16** (2012) 83-110.

[30] T. Haentjens & K. J. in 't Hout, *ADI schemes for pricing American options under the Heston model*, Appl. Math. Finan. **22** (2015) 207-237.

[31] E. Hairer, S. P. Nørsett & G. Wanner, Solving Ordinary Differential Equations I, 2nd ed., Springer, 1993.

[32] E. Hairer & G. Wanner, Solving Ordinary Differential Equations II, 2nd ed., Springer, 1996.

[33] Y. d'Halluin, P. A. Forsyth & K. R. Vetzal, *Robust numerical methods for contingent claims under jump diffusion processes*, IMA J. Numer. Anal. **25** (2005) 87-112.

[34] D. J. Higham, An Introduction to Financial Option Valuation, Cambridge Univ. Press, 2004.

[35] K. J. in 't Hout & S. Foulon, *ADI finite difference schemes for option pricing in the Heston model with correlation*, Int. J. Numer. Anal. Mod. **7** (2010) 303-320.

[36] K. J. in 't Hout & C. Mishra, *Stability of the modified Craig-Sneyd scheme for two-dimensional convection-diffusion equations with mixed-derivative term*, Math. Comp. Simul. **81** (2011) 2540-2548.

[37] K. J. in 't Hout & C. Mishra, *Stability of ADI schemes for multidimensional diffusion equations with mixed derivative terms*, Appl. Numer. Math. **74** (2013) 83-94.

[38] K. J. in 't Hout & J. Toivanen, *Application of operator splitting methods in finance*, in: "Splitting Methods in Communication, Imaging, Science, and Engineering", eds. R. Glowinski, S. J. Osher and W. Yin, Springer, 541-575 (2016).

[39] K. J. in 't Hout & K. Volders, *Stability of central finite difference schemes on non-uniform grids for the Black–Scholes equation*, Appl. Numer. Math. **59** (2009) 2593-2609.

[40] K. J. in 't Hout & K. Volders, *Stability and convergence analysis of discretizations of the Black–Scholes PDE with the linear boundary condition*, IMA J. Numer. Anal. **34** (2014) 296-325.

[41] K. J. in 't Hout & B. D. Welfert, *Stability of ADI schemes applied to convection-diffusion equations with mixed derivative terms*, Appl. Numer. Math. **57** (2007) 19-35.

[42] K. J. in 't Hout & B. D. Welfert, *Unconditional stability of second-order ADI schemes applied to multi-dimensional diffusion equations with mixed derivative terms*, Appl. Numer. Math. **59** (2009) 677-692.

[43] K. J. in 't Hout & M. Wyns, *Convergence of the Hundsdorfer-Verwer scheme for two-dimensional convection-diffusion equations with mixed derivative term*, AIP Conf. Proc. **1648** (2015) 850054.

[44] K. J. in 't Hout & M. Wyns, *Convergence of the Modified Craig-Sneyd scheme for two-dimensional convection-diffusion equations with mixed derivative term*, J. Comp. Appl. Math. **296** (2016) 170-180.

[45] S. D. Howison, C. Reisinger & J. H. Witte, *The effect of nonsmooth payoffs on the penalty approximation of American options*, SIAM J. Finan. Math. **4** (2013) 539-574.

[46] Y. Huang, P. A. Forsyth & G. Labahn, *Inexact arithmetic considerations for direct control and penalty methods: American options under jump diffusion*, Appl. Numer. Math. **72** (2013) 33-51.

[47] J. C. Hull, Options, Futures, and Other Derivatives, 9th ed., Pearson, 2014.

[48] W. Hundsdorfer, *Accuracy and stability of splitting with Stabilizing Corrections*, Appl. Numer. Math. **42** (2002) 213-233.

[49] W. Hundsdorfer & J. G. Verwer, Numerical Solution of Time-Dependent Advection-Diffusion-Reaction Equations, Springer, 2003.

[50] S. Ikonen & J. Toivanen, *Operator splitting methods for American option pricing*, Appl. Math. Lett. **17** (2004) 809-814.

[51] S. Ikonen & J. Toivanen, *Operator splitting methods for pricing American options under stochastic volatility*, Numer. Math. **113** (2009) 299-324.

[52] A. Itkin & P. Carr, *Jumps without tears: a new splitting technology for barrier options*, Int. J. Numer. Anal. Mod. **8** (2011) 667-704.

[53] P. Jaillet, D. Lamberton & B. Lapeyre, *Variational inequalities and the pricing of American options*, Acta Appl. Math. **21** (1990) 263-289.

[54] R. Kangro & R. Nicolaides, *Far field boundary conditions for Black–Scholes equations*, SIAM J. Numer. Anal. **38** (2000) 1357-1368.

[55] H. O. Kreiss, V. Thomée & O. Widlund, *Smoothing of initial data and rates of convergence for parabolic difference equations*, Comm. Pure Appl. Math. **23** (1970) 241–259.

[56] Y. Kwon & Y. Lee, *A second-order finite difference method for option pricing under jump-diffusion models*, SIAM J. Numer. Anal. **49** (2011) 2598–2617.

[57] P. Leoni, The Greeks and Hedging Explained, Palgrave Macmillan, 2014.

[58] A. Lipton, Mathematical Methods for Foreign Exchange, World Scientific, 2001.

[59] X. Mao, Stochastic Differential Equations and Applications, 2nd ed., Horwood, 2008.

[60] S. McKee & A. R. Mitchell, *Alternating direction methods for parabolic equations in two space dimensions with a mixed derivative*, Computer J. **13** (1970) 81–86.

[61] S. McKee, D. P. Wall & S. K. Wilson, *An alternating direction implicit scheme for parabolic equations with mixed derivative and convective terms*, J. Comp. Phys. **126** (1996) 64–76.

[62] R. C. Merton, *Theory of rational option pricing*, Bell J. Econ. Manag. Sc. **4** (1973) 141–183.

[63] R. C. Merton, *Option pricing when underlying stock returns are discontinuous*, J. Finan. Econ. **3** (1976) 125–144.

[64] C. Mishra, *A new stability result for the modified Craig-Sneyd scheme applied to two-dimensional convection-diffusion equations with mixed derivatives*, Appl. Math. Comp. **285** (2016) 41–50.

[65] A. R. Mitchell & D. F. Griffiths, The Finite Difference Method in Partial Differential Equations, Wiley, 1980.

[66] S. N. Neftci, An Introduction to the Mathematics of Financial Derivatives, 2nd ed., Academic Press, 2000.

[67] G. G. O'Brien, M. A. Hyman & S. Kaplan, *A study of the numerical solution of partial differential equations*, J. Math. and Phys. **29** (1951) 223–251.

[68] B. Øksendal, Stochastic Differential Equations, 5th ed., Springer, 2010.

[69] D. W. Peaceman & H. H. Rachford, *The numerical solution of parabolic and elliptic differential equations*, J. Soc. Ind. Appl. Math. **3** (1955) 28–41.

[70] D. M. Pooley, P. A. Forsyth & K. R. Vetzal, *Numerical convergence properties of option pricing PDEs with uncertain volatility*, IMA J. Numer. Anal. **23** (2003) 241–267.

[71] D. M. Pooley, K. R. Vetzal & P. A. Forsyth, *Convergence remedies for non-smooth payoffs in option pricing*, J. Comp. Finan. **6** (2003) 25–40.

[72] R. Rannacher, *Finite element solution of diffusion problems with irregular data*, Numer. Math. **43** (1984) 309–327.

[73] C. Reisinger & A. Whitley, *The impact of a natural time change on the convergence of the Crank–Nicolson scheme*, IMA J. Numer. Anal. **34** (2014) 1156–1192.

[74] R. D. Richtmyer & K. W. Morton, Difference Methods for Initial-Value Problems, 2nd ed., Wiley, 1967.

[75] S. Salmi & J. Toivanen, *IMEX schemes for pricing options under jump-diffusion models*, Appl. Numer. Math. **84** (2014) 33–45.

[76] S. Salmi, J. Toivanen & L. von Sydow, *An IMEX-scheme for pricing options under stochastic volatility models with jumps*, SIAM J. Sci. Comp. **36** (2014) B817–B834.

[77] W. Schoutens, Lévy Processes in Finance, Wiley, 2003.

[78] E. S. Schwartz, *The valuation of warrants: implementing a new approach*, J. Finan. Econ. **4** (1977) 79-93.

[79] R. U. Seydel, Tools for Computational Finance, 5th ed., Springer, 2012.

[80] S. E. Shreve, Stochastic Calculus for Finance II, 8th pr., Springer, 2008.

[81] G. D. Smith, Numerical Solution of Partial Differential Equations: Finite Difference Methods, 3rd ed., Clarendon Press, 1985.

[82] M. N. Spijker, Numerical Stability, Lecture notes, Univ. Leiden, 1998.

[83] J. C. Strikwerda, Finite Difference Schemes and Partial Differential Equations, Wadsworth, 1989.

[84] R. M. Stulz, *Options on the minimum or the maximum of two risky assets*, J. Finan. Econ. **10** (1982) 161-185.

[85] D. Tavella & C. Randall, Pricing Financial Instruments, Wiley, 2000.

[86] J. W. Thomas, Numerical Partial Differential Equations: Finite Difference Methods, Springer, 1995.

[87] L. N. Trefethen & M. Embree, Spectra and Pseudospectra, Princeton Univ. Press, 2005.

[88] A. E. P. Veldman & K. Rinzema, *Playing with nonuniform grids*, J. Eng. Math. **26** (1992) 119-130.

[89] B. A. Wade, A. Q. M. Khaliq, M. Yousuf, J. Vigo-Aguiar & R. Deininger, *On smoothing of the Crank-Nicolson scheme and higher order schemes for pricing barrier options*, J. Comp. Appl. Math. **204** (2007) 144-158.

[90] P. Wilmott, Derivatives, Wiley, 1999.

[91] P. Wilmott, J. Dewynne & S. Howison, Option Pricing, Oxford Financial Press, 1993.

[92] H. Windcliff, P. A. Forsyth & K. R. Vetzal, *Analysis of the stability of the linear boundary condition for the Black-Scholes equation*, J. Comp. Finan. **8** (2004) 65-92.

[93] M. Wyns, *Convergence analysis of the Modified Craig-Sneyd scheme for two-dimensional convection-diffusion equations with nonsmooth initial data*, published online in IMA J. Numer. Anal. (2016), doi: 10.1093/imanum/drw028

[94] Y. I. Zhu, X. Wu, I. L. Chern & Z. z. Sun, Derivative Securities and Difference Methods, 2nd ed., Springer, 2013.

[95] R. Zvan, P. A. Forsyth & K. R. Vetzal, *Penalty methods for American options with stochastic volatility*, J. Comp. Appl. Math. **91** (1998) 199-218.

[96] R. Zvan, P. A. Forsyth & K. R. Vetzal, *Robust numerical methods for PDE models of Asian options*, J. Comp. Finan. **1** (1998) 39-78.

Index

© The Author(s) 2017   **127**
K. in 't Hout, *Numerical Partial Differential Equations in Finance Explained*,
Financial Engineering Explained, DOI 10.1057/978-1-137-43569-9